AF821049

Comparison Theorems in Riemannian Geometry

Photo courtesy of Ann Billingsley

David G. Ebin and Jeff Cheeger

COMPARISON THEOREMS IN RIEMANNIAN GEOMETRY

JEFF CHEEGER
DAVID G. EBIN

AMS CHELSEA PUBLISHING

AMERICAN MATHEMATICAL SOCIETY
Providence, Rhode Island

2000 *Mathematics Subject Classification.* Primary 53C20; Secondary 58E10.

For additional information and updates on this book, visit
www.ams.org/bookpages/chel-365

Library of Congress Cataloging-in-Publication Data

Cheeger, Jeff.
Comparison theorems in riemannian geometry / Jeff Cheeger, David G. Ebin.
p. cm. — (AMS Chelsea Publishing)
Originally published: Amsterdam : North-Holland Pub. Co. ; New York : American Elsevier Pub. Co., 1975, in series: North-Holland mathematical library ; v. 9.
Includes bibliographical references and index.
ISBN 978-0-8218-4417-5 (alk. paper) | Softcover ISBN 978-1-4704-8111-7
1. Geometry, Riemannian. 2. Riemannian manifolds. I. Ebin, D. G. II. American Mathematical Society. III. Title.

QA649.C47 2008
516.3′73—dc22 2007052113

Copying and reprinting. Individual readers of this publication, and nonprofit libraries acting for them, are permitted to make fair use of the material, such as to copy a chapter for use in teaching or research. Permission is granted to quote brief passages from this publication in reviews, provided the customary acknowledgment of the source is given.

Republication, systematic copying, or multiple reproduction of any material in this publication is permitted only under license from the American Mathematical Society. Requests for such permission should be addressed to the Acquisitions Department, American Mathematical Society, 201 Charles Street, Providence, Rhode Island 02904-2294, USA. Requests can also be made by e-mail to `reprint-permission@ams.org`.

© 1975 held by the American Mathematical Society. All rights reserved.
Reprinted by the American Mathematical Society, 2008, 2025.
Printed in the United States of America.

∞ The paper used in this book is acid-free and falls within the guidelines
established to ensure permanence and durability.
Visit the AMS home page at `http://www.ams.org/`

12 11 10 9 8 7 6 5 4 3   30 29 28 27 26 25

Contents

Preface to the AMS Chelsea Printing

In the period since the first edition of this book appeared (1976), Riemannian geometry has experienced explosive growth. This circumstance was ultimately the determining factor in our decision not to try to bring the book up to date. We came to realize that if such an attempt were made, the book would never see the light of day. So our book remains essentially as it was. We hope that it will continue to play a useful role by setting out some (still relevant) fundamentals of the subject.

There do exist a number of valuable surveys and expositions of many of the developments that have taken place in the intervening years. Some of these are indicated in a relatively small collection of additional references that have been added to the bibliography. The surveys of Berger, in particular, provide a sweeping overview and include very extensive bibliographies.

We have also added a highly selective (and quite inadequate) set of references to original sources, reflecting what we see as some of the most significant developments over the past 30 years. We apologize in advance to anyone whose work has been unjustly slighted.

We are indebted to Satyaki Dutta for having retyped the manuscript and for a substantial amount of proofreading. We would also like to express our thanks to all those colleagues who, in the 20 years since the first edition went out of print, have urged us to consider having it republished.

Jeff Cheeger and David G. Ebin, 2007.

Preface

In this book we study complete Riemannian manifolds by developing techniques for comparing the geometry of a general manifold M with that of a simply connected model space of constant curvature M_H. A typical conclusion is that M retains particular geometrical properties of the model space under the assumption that its sectional curvature K_M, is bounded between suitable constants. Once this has been established, it is usually possible to conclude that M retains topological properties of M_H as well.

The distinction between strict and weak bounds on K_M is important, since this may reflect the difference between the geometry of say the sphere and that of Euclidean space. However, it is often the case that a conclusion which becomes false when one relaxes the condition of strict inequality to weak inequality can be shown to fail only under very special circumstances. Results of this nature, which are known as rigidity theorems, generally require a delicate global argument. Here are some examples which will be treated in more detail in Chapter 8.

Topological Theorem. If M is a complete manifold such that $K_M \geq \delta > 0$, then M has finite fundamental group.

Geometrical Antecedent. If M is a complete manifold such that $K_M \geq \delta > 0$, then the diameter of its universal covering space $\widetilde{M}$ is $\leq \Pi/\sqrt{\delta}$. In particular, $\widetilde{M}$ is compact.

Even if we assume M to be compact, the preceding statements are false if only $K_M \geq 0$. However, we can show the following.

Rigid Topological Theorem. Let M be a compact manifold such that $K_M \geq 0$. Then there is an exact sequence

$$0 \to \Phi \to \Pi_1(M) \to B \to 0$$

where is a finite group and B is a crystallographic group on $\mathbb{R}_k$ for some $k \leq \dim M$, and therefore satisfies an exact sequence

$$0 \to \mathbb{Z}_k \to B \to \Psi \to 0$$

where Ψ is a finite group.

Rigid Geometrical Antecedent. Let M be a compact manifold such that $K_M \geq 0$. Then $\widetilde{M}$ splits isometrically as $\overline{M} \times \mathbb{R}_k$ (same k as above), where $\overline{M}$ is compact and $\mathbb{R}_k$ has its standard flat metric. Thus, if $K_M \geq 0$, $\widetilde{M}$ may not be compact, but it is at worst the isometric product of a compact manifold and Euclidean space. The infinite part of $\Pi_1(M)$ comes precisely from the Euclidean factor.

The reader of this book should have a basic knowledge of differential geometry and algebraic topology, at least the equivalent of a one term course is each. Our purpose is to provide him with a fairly direct route to some interesting geometrical theorems, without his becoming bogged down in a detailed study of connections and tensors. In keeping with this approach, we have limited ourselves primarily to those techniques which arise as outgrowths of the second variation formula and to some extent of Morse theory.

In Chapter 1 we have included a rapid treatment of the more elementary material on which the later chapters are based. Of course, we do not recommend that the less knowledgeable reader regard this as a comprehensive introduction to Riemannian geometry. Likewise, Chapters 3 and 4 are provided in part for the convenience of the reader. In Chapter 3 which deals with homogeneous spaces, we also summarize without proof the relevant material on Lie groups. In Chapter 4 the main theorems of Morse theory are stated, again mostly without proofs. An exception, however, is Lemma 4.11, which is perhaps less standard than the other material. For the unproven results in both chapters, excellent references are readily available. Our main geometrical tools, the Rauch Comparison Theorems and the more global Toponogov Theorem, are discussed in Chapters 1 and 2 respectively. Chapter 5 deals with closed geodesics and the injectivity radius of the exponential map. Chapters 6-9 form the core of our study. Chapter 6 contains the Sphere Theorem – M simply connected and $1 \geq K_M > 1/4$ implies M homeomorphic to a sphere – as well as Berger's rigidity theorem which covers the case $1 \geq K_M \geq 1/4$. The last three chapters deal with material of recent origin. Chapter 7 is primarily concerned with the differentiable version of the Sphere Theorem. Chapter 8 takes up the structure theory of complete noncompact manifolds of nonnegative curvature, while Chapter 9 gives some results on the fundamental group of compact manifolds of nonpositive curvature.

It is a pleasure to thank Carole Alberghine and Lois Cheeger for their patient work, typing the original manuscript.

CHAPTER 1

Basic Concepts and Results

1. Notation and preliminaries

We will begin by fixing some notation and recalling some standard facts about connections. M will denote a smooth (i.e. C^∞) finite-dimensional manifold and $T(M)$ its tangent bundle. Sometimes we write M^n to indicate that M has dimension n. For basic concepts about manifolds, such as the notions of submanifolds, immersion, embedding and diffeomorphism, we refer to Kobayashi and Nomizu [1963, 1969]. For $p \in M$, M_p denotes the tangent space to M at p. $\chi(M)$ or χ is the linear space of smooth vector fields on M and $\mathfrak{F}(M)$ the ring of smooth functions on M. We will usually denote a tangent vector at a point by a lower case letter and its extension to a vector field by the corresponding capital.

A Riemannian metric is an assignment to each $p \in M$ of a symmetric positive-definite bilinear form $\langle\ ,\ \rangle_p$ on M_p such that for any $V, W \in \chi(M)$ the function $p \to \langle V, W\rangle_p$ is in $\mathfrak{F}(M)$. Also, $\langle V, V\rangle^{\frac{1}{2}}$ is denoted by $||V||$.

An *affine connection* is a bilinear map

$$\nabla : \chi(M) \times \chi(M) \to \chi(M),$$

which has the following properties:

$$\nabla_{fV} W = f\nabla_V W, \tag{1.1}$$

$$\nabla_V fW = (Vf)W + f\nabla_V W \tag{1.2}$$

for any $f \in \mathfrak{F}(M)$, $V, W \in \chi(M)$.

We call $\nabla_V W$ the *covariant derivative* of W in the direction of V. We refer to Bishop and Crittenden [1964] and Kobayashi and Nomizu [1963, 1969] for more detailed versions of connection theory.

The Fundamental Theorem of Riemannian Geometry states that for each Riemannian metric there is a unique affine connection, called the *Riemannian connection*, with the following two properties:

$$X\langle V, W\rangle = \langle \nabla_X V, W\rangle + \langle V, \nabla_X W\rangle, \tag{1.3}$$

$$\nabla_V W - \nabla_W V - [V, W] = 0, \tag{1.4}$$

where $[\ ,\ ]$ signifies the Lie bracket, $[X, Y]f = (XY - YX)f$. The first of these properties is a condition of compatibility between the affine connection and the metric, while the second is a symmetry condition on the connection alone. In general, the quantity set equal to zero in (1.4) is called the *torsion* of the connection, $\mathrm{Tor}(v, w)$. It is a tensor of type (1,2). Hence the Fundamental Theorem may be paraphrased by saying that there is a unique torsion-free connection compatible with any given metric. We recall the proof.

To show uniqueness, it suffices to show that $\langle \nabla_X Y, Z\rangle$ is determined by (1.3), (1.4). Using (1.4),

$$\begin{aligned} X\langle Y, Z\rangle &= \langle \nabla_X Y, Z\rangle + \langle Y, \nabla_X Z\rangle, \\ Y\langle X, Z\rangle &= \langle \nabla_Y X, Z\rangle + \langle X, \nabla_Y Z\rangle, \\ Z\langle X, Y\rangle &= \langle \nabla_Z X, Y\rangle + \langle X, \nabla_Z Y\rangle, \end{aligned}$$

Subtracting the third of these equations from the sum of the first two and using (1.4) yields

$$\begin{aligned} 2\langle \nabla_X Y, Z\rangle = X\langle Y, Z\rangle &+ Y\langle X, Z\rangle - Z\langle X, Y\rangle \\ &+ \langle [X, Y], Z\rangle - \langle [X, Z], Y\rangle - \langle [Y, Z], X\rangle. \end{aligned}$$

Conversely, if we use this formula to define $\langle \nabla_X Y, Z\rangle$, it is straightforward to check that we get a connection satisfying (1.3) and (1.4).

From (1.1), (1.2) it is easy to see that $(\nabla_V W)(p)$ is determined by W and $V(p)$. In fact it is not hard to show that it is determined by $V(p)$ and W restricted to any curve through p in direction $V(p)$. If α is a 1-form, we define $\nabla_V \alpha$ by the equation

$$(\nabla_V \alpha)(W) + \alpha(\nabla_V W) = V(\alpha(W)).$$

We then define $\nabla_V \sigma$ for any tensor field σ by extending ∇ as a derivation.

Let $c : [0, 1] \to M$ be a smooth curve, and let $c'(t)$ denote the tangent vector to $c(t)$. For any $v \in M_{c(0)}$ there is a unique vector $V(t) \in M_{c(t)}$ such that $V(0) = v$ and $\nabla_{c'(t)} V(t) \equiv 0$ We call $V(t)$ the *parallel field* and $V(1)$ the *parallel translate* of v along $c(t)$. In fact, one can easily find smooth vector fields along $c(t)$, $E_1(t), \dots, E_n(t)$ such that $\{E_i(t)\}$ is an orthonormal basis for $M_{c(t)}$. In terms of these fields, the above equation may be written as a first-order system of ordinary differential equations. We have

$$\frac{\mathrm{d}}{\mathrm{d}t}\langle V, E_i\rangle = c'\langle V, E_i\rangle = \langle \nabla_{c'} V, E_i\rangle + \langle V, \nabla_{c'} E_i\rangle = \langle \nabla_{c'} E_i, V\rangle.$$

Therefore

$$\begin{bmatrix} \langle V, E_1\rangle \\ \vdots \\ \langle V, E_n\rangle \end{bmatrix}' = \langle \nabla_{c'} E_i, E_j\rangle \begin{bmatrix} \langle V, E_1\rangle \\ \vdots \\ \langle V, E_n\rangle \end{bmatrix}.$$

The existence theorem for ordinary differential equations tells us that we can solve for $V(t)$. Because the above equation is linear, we obtain a linear map $P_c : M_{c(0)} \to M_{c(1)}$ defined by $v \to V(1)$. It follows easily from (1.3) that P_c is an isometry.

In the sequel we will often be confronted with the following situation: Let $\phi : N \to M$ be a smooth map, and let M have a connection ∇. A *vector field along* ϕ is an assignment $x \to W$, where $W \in M_{\phi(x)}$. Let $\{E_i\}$ be a frame field in a neighborhood of $\phi(x)$. Then we can write

$$W(x) = \Sigma f_i(x) E_i(\phi(x)).$$

We say that $W(x)$ is *smooth* if the functions $f_i(x)$ are smooth. If $v \in N_x$, we can define $\tilde{\nabla}_v W \in M_{\phi(x)}$, the covariant derivative of W in the direction of v, by

$$\tilde{\nabla}_v W = \Sigma v(f_i(x)) E_i(\phi(x)) + f_i(x) \nabla_{\mathrm{d}\phi(v)} E_i(\phi(x)).$$

The definition is easily seen to be independent of the choice of $\{E_i(x)\}$. Let M be Riemannian and ∇ the Riemannian connection. If $v \in N_x$, W_1, W_2 are vector fields along ϕ, then one easily checks that

$$v\langle W_1, W_2\rangle = \langle \tilde{\nabla}_v W_1, W_2\rangle + \langle W_1, \tilde{\nabla}_v W_2\rangle. \qquad (*)$$

Also, if V_1, V_2 are vector fields in N, then $\mathrm{d}\phi(V_1)$, $\mathrm{d}\phi(V_2)$ are vector fields along ϕ and

$$\tilde{\nabla}_{V_1} \mathrm{d}\phi(V_2) - \tilde{\nabla}_{V_2} \mathrm{d}\phi(V_1) - \mathrm{d}\phi([V_1, V_2]) = 0. \qquad (**)$$

A vector field along ϕ is also called a *section* of the induced bundle $\phi^*(TM)$. We call $\tilde{\nabla}$ the *induced connection*. In the proofs of the first and second variation formulas and elsewhere, implicitly we will be using (*) and (**). However, for convenience we will suppress the notation $\tilde{\nabla}$ and proceed formally as if the vector fields along ϕ were actually defined on M.

2. First variation of arc length

Let M be a Riemannian manifold. We denote the length of the continuous piecewise smooth curve $c : [a, b] \to M$ by $L[c]$. By definition,

$$L[c] = \int_a^b ||c'(t)|| \mathrm{d}t.$$

It follows from the chain rule that $L[c]$ does not depend on a particular choice of parameterization. M becomes a metric space if we define the distance between two points as the infimum of the lengths of all curves between them. We denote the distance from p to q by $\rho(p, q)$. When proving that M is a metric space, the only point which is not entirely trivial to check is that if $p \neq q$, then $\rho(p, q) > 0$. To see this, let $x_1, \ldots, x_n$ be a local coordinate system with p at the origin, and let $B_r(p)^-$ denote the set $\Sigma x_i^2 \leq r$. Let g denote the given Riemannian metric and $\bar{g}$ the Euclidean metric $\bar{g}(\frac{\partial}{\partial x_i}, \frac{\partial}{\partial x_j}) = \delta_{ij}$. At each point, g may be diagonalized with respect to $\bar{g}$, and the smallest eigenvalue $\lambda(x)$ is a positive continuous function on $B_r(p)^-$. It follows easily that any curve from p to the boundary of $B_r(p)^-$ has length at least $\frac{r}{\lambda_0}$, where λ_0 is the minimum of $\lambda(x)$. However, given $q \neq p$, we may assume that r has been chosen sufficiently small so that $q \notin B_r(p)^-$. Then any curve from p to q must contain an initial segment from p to the boundary of $B_r(p)^-$. Therefore $\rho(p, q) > \frac{r}{\lambda_0}$. A similar argument shows that the metric space and manifold topologies coincide.

Let $p, q \in M$. In the next three sections we will investigate the conditions under which there exists a curve γ from p to q such that $L[\gamma] = \rho(p,q)$. We are now going to derive a necessary condition for the existence of such a curve. We begin by assuming that c is smooth, $||c'(t)|| \neq 0$, and without loss of generality we may assume c to be parameterized proportional to arc length. In other words, $||c'(t)||$ is a constant l.

Let $\alpha : Q \to M$ be a smooth function, where Q is the rectangle $[a,b] \times (-\epsilon, \epsilon)$ and such that

$$\alpha|[a,b] \times \{0\} = c : [a,b] \to M.$$

α is said to define a *smooth variation* of the curve $\alpha|[a,b] \times \{0\}$. Let T, V be the fields of tangent vectors on Q corresponding to its first and second variables. We shall identify these vectors with their images under the differential of α. Our goal is to compute the change in arc length over the family of curves $c_s = \alpha|[a,b] \times \{s\}$, where $-\epsilon < s < \epsilon$. This is given by

$$\begin{aligned}\frac{\mathrm{d}}{\mathrm{d}s} L[c_s] &= \frac{\mathrm{d}}{\mathrm{d}s} \int_a^b \langle c_s'(t), c_s'(t)\rangle^{\frac{1}{2}} \mathrm{d}t = \int_a^b V\langle T, T\rangle^{\frac{1}{2}} \mathrm{d}t \\ &= \frac{1}{2}\int_a^b \langle T, T\rangle^{-\frac{1}{2}} V\langle T, T\rangle \mathrm{d}t = \int_a^b \langle T, T\rangle^{-\frac{1}{2}} \langle \nabla_V T, T\rangle \mathrm{d}t.\end{aligned}$$

Since $[T, V] = 0$ on Q, using (**) we may rewrite this as

$$\int_a^b \langle T, T\rangle^{-1/2} \langle \nabla_T V, T\rangle \mathrm{d}t.$$

Since $||c_0'|| = l$,

$$\begin{aligned}\frac{\mathrm{d}}{\mathrm{d}s} L[c_s]|_{s=0} &= l^{-1} \int_a^b \langle \nabla_T V, T\rangle \mathrm{d}t \\ &= l^{-1} \int_a^b (T\langle V, T\rangle - \langle V, \nabla_T T\rangle \mathrm{d}t.\end{aligned}$$

By integrating the first term we obtain

$$\frac{\mathrm{d}}{\mathrm{d}s} L[c_s]|_{s=0} = l^{-1}\Big\{\langle V, T\rangle|_a^b - \int_a^b \langle V, \nabla_T T\rangle \mathrm{d}t\Big\}. \tag{1.5}$$

This expression is called the *first variation formula*.

Suppose that rather than being smooth, the function α above is continuous, and has the property that $a = t_0 < t_1 < \cdots < t_n = b$ is some subdivision of $[a,b]$ such that $\alpha_{[t_i, t_{i+1}] \times (-\epsilon, \epsilon)}$ is smooth. In this case, α is said to define a *piecewise smooth* variation. The first variation formula for piecewise smooth variations is obtained by applying (1.5) to each segment $[t_i, t_{i+1}]$. In particular, if c_0 is actually smooth, the intermediate terms cancel, and (1.5) remains valid. A vector field V along c_0 is called piecewise smooth if V is continuous and there exists a subdivision as above such that $V|[t_i, t_{i+1}]$ is smooth vector field along c_0. It is important to remark that any piecewise smooth V arises from some variation. In fact, for sufficiently small s, the variation $\alpha : (t,s) \to \exp_{c_0(t)}(sV)$ will suffice. (See Section 3 for the definition of exp).

If all curves c_s have the same endpoints, then $V(a,0) = V(b,0) = 0$, so that

$$\frac{\mathrm{d}}{\mathrm{d}s} L[c_s]|_{s=0} = -l^{-1} \int_a^b \langle V, \nabla_T T \rangle \mathrm{d}t.$$

If $c = c_0$ is the shortest curve from $c(a)$ to $c(b)$, then

$$\frac{\mathrm{d}}{\mathrm{d}s} L[c_s]|_{s=0} = 0$$

for any map $\alpha : Q \to M$. Hence for V any vector field along c which vanishes at the endpoints, the right-hand side of (1.5) vanishes. Therefore, if c is smooth and minimal, by taking a variation such that $V = \phi(t)\nabla_T T$ for some function $\phi(t)$ such that $\phi(t) > 0$ for $a < t < b$ and $\phi(a) = \phi(b) = 0$, we conclude that

$$\nabla_T T = \nabla_{c'} c' \equiv 0.$$

The preceding calculation motivates the following definition.

DEFINITION 1.6. We call a smooth curve c a *geodesic* if $\nabla_{c'} c' \equiv 0$.

If c is a geodesic,

$$c'\langle c', c' \rangle = 2\langle \nabla_{c'} c', c' \rangle = 0,$$

so $\langle c', c' \rangle$ must be constant. Thus a geodesic is parameterized proportional to arc length, and by (1.5) it is always a critical point of the arc-length function under any variation with fixed end points. In fact, for this we only need to know that the variation is perpendicular to T at the endpoints, or, more generally, that $\langle T, V \rangle|_a^b = 0$. If $||c'|| = 1$, c is called a *normal* geodesic.

The following useful proposition illustrates how the first variation formula may be applied to obtain geometrical information.

PROPOSITION 1.7. *Let N and $\bar{N}$ be two submanifolds of M, without boundary, and let $\gamma : [0,t] \to M$ be a geodesic such that $\gamma(0) \in N$, $\gamma(t) \in \bar{N}$ and γ is the shortest curve from N to $\bar{N}$. Then $\gamma'(0)$ is perpendicular to $N_{\gamma(0)}$ and $\gamma'(t)$ is perpendicular to $\bar{N}_{\gamma(t)}$*

PROOF. 1If $\gamma'(0)$ is not perpendicular to $N_{\gamma(0)}$, choose $x \in N_{\gamma(0)}$ such that $\langle \gamma'(0), x \rangle > 0$, and let c be a curve in N starting at $\gamma(0)$ such that $c'(0) = x$.

Construct a variation $\alpha : [0,t] \times (-\epsilon, \epsilon) \to M$ such that

$$\alpha|[0,t] \times \{0\} = \gamma, \qquad \alpha(0,s) = c(s), \alpha(t,s) = \gamma(t).$$

Then if $\gamma_s = \alpha|[0,t] \times \{s\}$, formula (1.5) shows that

$$\frac{\mathrm{d}}{\mathrm{d}s} L[\gamma_s]|_{s=0} = -l^{-1}\langle \gamma'(0), x \rangle < 0.$$

Therefore, for small s, $L[\gamma_s] < L[\gamma]$, and γ is not minimal. A completely analogous argument shows that $\gamma'(t)$ must also be perpendicular to $\bar{N}_{\gamma(t)}$. (See Fig. 1.1) $\square$

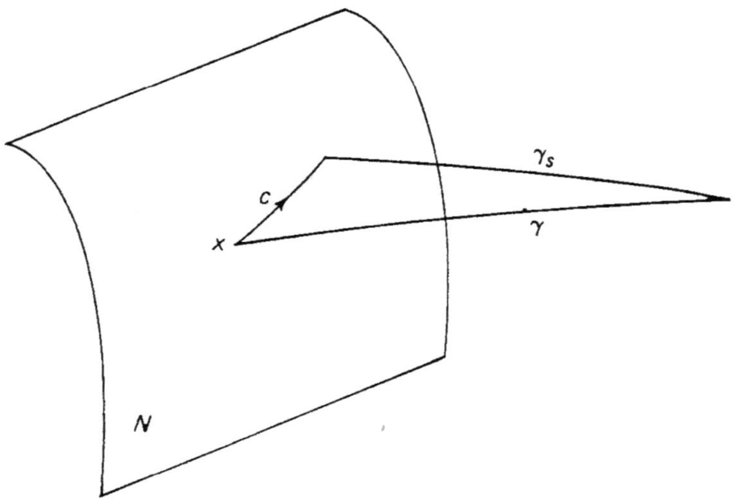

Fig. 1.1.

Note that in the above variation it is not necessary that the curves γ_s be geodesics. As long as α is a smooth map, our calculation shows that for small s, $L[\gamma_s] < L[\gamma]$.

3. Exponential map and normal coordinates

A fundamental property of geodesics is that, given a point $p \in M$ and a vector $v \in M_p$, there exists a unique geodesic γ_v through p whose tangent at p is v. This follows from the fact that the defining condition for geodesics, $\nabla_{\gamma'}\gamma' = 0$, is a second-order differential equation in the parameter t of γ. p and v are exactly the required initial conditions for existence and uniqueness of γ_v (see Milnor [1963], p.56, for details). If $\gamma_v : (-\epsilon, \epsilon) \to M$ is a geodesic parameterized by t, the curve $c : (-\frac{\epsilon}{s}, \frac{\epsilon}{s}) \to M$ defined by $c(t) = \gamma_v(st)$ (s fixed) is also a geodesic; for $\nabla_{c'}c' = s^2\nabla_{\gamma_v}\gamma_v = 0$. Also $c'(0) = sv$, so $c = \gamma_{sv}$.

The *exponential map* $\exp_p : M_p \to M$ is defined by $\exp_p(v) = \gamma_v(1)$ for all $v \in M_p$ such that 1 is in the domain of γ_v. From the above we know that for any fixed v there exists a number $s > 0$ such that $\gamma_{sv}(1)$ is defined, and by the existence theorem of second-order differential equations, we can pick s to vary continuously with v. It follows that $\exp_p$ is defined on a neighborhood of the origin in M_p. Furthermore, it is a smooth map and, by the implicit function theorem, a local diffeomorphism in a neighborhood of the origin. Actually, since we have $\exp_p : M_p \to M$ for each $p \in M$, we can define the union of these maps $\exp : T(M) \to M$. We shall define exp on the union over p of the domains of $\exp_p$, which is a neighborhood of the zero section of $T(M)$.

If we choose an orthonormal basis $\{e_i\}$ for M_p, we can define a coordinate system in a neighborhood of p by assigning to the point $\exp_p(\Sigma x^i e_i)$ the coordinates $(x_1, \ldots, x_n)$. Such coordinates are called *normal coordinates* at p. Since the rays through the origin are geodesics, normal coordinates have the property that $\nabla_{\frac{\partial}{\partial x_i}}(\frac{\partial}{\partial x_i}) \equiv 0$. It follows that for all v in M_p,

$$\left(\nabla_v \frac{\partial}{\partial x_i}\right)\Big|_p = 0.$$

For these reasons, normal coordinates are convenient to use.

If M is a Riemannian manifold, by definition each tangent space M_p comes equipped with an inner product. For each $v \in M_p$, the tangent space $(M_p)_v$ can be naturally identified with M_p. Hence $(M_p)_v$ inherits an inner product. Fixing $v \in M_p$, $\mathrm{dexp} : (M_p)_v \to M_{\exp(v)}$ is a linear map. dexp does not in general preserve the inner products on these spaces.

If we let $\rho(t) = tv$ be the ray from $0 \in M_p$ through v, and assume that exp is defined along ρ, it is easy to verify that $\mathrm{dexp}(\rho'(t)) = \gamma_v'(t)$ and that $||\rho'(t)|| = ||\gamma_v'(t)||$. Moreover, we have the following important result known as the Gauss Lemma.

GAUSS LEMMA 1.8. *If $\rho(t) = tv$ is a ray through the origin of M_p and $w \in (M_p)_{\rho(t)}$ is perpendicular to $\rho'(t)$, then* $\mathrm{dexp}(w)$ *is perpendicular to* $\mathrm{dexp}(\rho'(t))$.

PROOF. Let $c(s)$ be a curve in M_p such that $c(0) = v$, $c'(0) = w$ and such that every point of c is at the same distance from the origin of M_p. Let $\alpha(t, s)$ be a rectangle in M defined by

$$\alpha(t, s) = \exp(\rho_s(t)),$$

where $\rho_s : [0, 1] \to M_p$ is the ray from 0 to $c(s)$ in M_p. From the definition of exp and α we know that the lengths of the curves $t \to \alpha(t, s)$ are independent of s. Also, $t \to \alpha(t, 0)$ is a geodesic, and

$$\mathrm{d}\alpha\frac{\partial}{\partial s}(0, 0) = 0, \qquad \mathrm{d}\alpha\frac{\partial}{\partial s}(0, 1) = \mathrm{dexp}(w).$$

The first variation formula gives

$$0 = \frac{\mathrm{d}}{\mathrm{d}s}L[\exp(\rho_s)]\Big|_{s=0} = \langle \mathrm{dexp}(w), \gamma_v' \rangle|_0^t = \langle \mathrm{dexp}(w), \gamma_v'(t) \rangle.$$

The lemma follows. □

The Gauss Lemma is equivalent to the fact that on a normal coordinate ball with the origin deleted the *gradient* of the function $r = (\Sigma x_i^2)^{\frac{1}{2}}$ is $\frac{\partial}{\partial r}$. Recall that the gradient of a function f is the unique vector field defined by

$$\langle \mathrm{grad} f, x \rangle = \mathrm{d}f(x) = x(f).$$

In fact, using polar coordinates $r, \theta_1, \dots, \theta_{n-1}$ (which are defined like normal coordinates above) the Gauss Lemma implies that $\langle \frac{\partial}{\partial \theta_i}, \frac{\partial}{\partial r} \rangle = 0$. Then if

$$x = h\frac{\partial}{\partial r} + \sum g_i \frac{\partial}{\partial \theta_i},$$

we have

$$\Big\langle \frac{\partial}{\partial r}, x \Big\rangle = h = h\frac{\partial r}{\partial r} + \sum g_i \frac{\partial r}{\partial \theta_i} = x(r).$$

So far we have not shown, even for sufficiently small ϵ, that the length of a normal geodesic segment $\gamma : [0, \epsilon] \to M$ equals the distance between its endpoints. The following corollary of the Gauss Lemma shows that this is indeed the case. In fact we show directly that γ has shorter length than any other curve between its endpoints. This suffices because distance is defined as the infimum of lengths of such curves.

COROLLARY 1.9. *Let $B_r(0) \subseteq M_p$ be a ball of radius r on which $\exp_p$ is an embedding. Then:*

(1) For $v \in B_r(0)$, $\gamma_v : [0, 1] \to M$ is the unique curve satisfying

$$L[\gamma] = \rho(p, \exp_p(v)) = ||v||.$$

In particular for any curve c, if $L[c] = \rho(c(0), c(1))$, then, up to reparameterization, c is a smooth geodesic.

(2) If $q \notin \exp_p(B_r(0)) = B_r(p)$, then there exists $q' \in \partial B_r(p)$, the boundary of $B_r(p)$, such that $\rho(p, q) = r + \rho(q', q)$. In particular, $\rho(p, q) \geq r$.

PROOF. (1) Let $c : [0, 1] \to M$ be a piecewise smooth curve from p to $\exp_p(v)$. Assume $c(t) \in \exp_p(B_r(0))$ for $t \leq t_0$, i.e. $r(c(t)) \leq r$ for $t \leq t_0$. Since $||\frac{\partial}{\partial r}|| \equiv 1$, it follows that where $c(t)$ is smooth we have

$$||c'(t)|| \geq \Big\langle c'(t), \frac{\partial}{\partial r} \Big\rangle,$$

with equality holding if and only if $c'(t) = \lambda(t)(\frac{\partial}{\partial r})$, with $\lambda(t) \geq 0$. Then

$$\begin{aligned} L[c] = \int_0^1 ||c'|| \mathrm{d}t &= \int_0^{t_0} ||c'|| \mathrm{d}t + \int_{t_0}^1 ||c'|| \mathrm{d}t \\ &\geq \int_0^{t_0} \Big\langle c', \frac{\partial}{\partial r} \Big\rangle \mathrm{d}t + \int_{t_0}^1 ||c'|| \mathrm{d}t. \end{aligned}$$

As we have seen, grad $r = \frac{\partial}{\partial r}$. Thus the right-hand side of the above is equal to

$$\int_0^{t_0} \frac{\mathrm{d}}{\mathrm{d}t} r(c(t)) \mathrm{d}t + \int_{t_0}^1 ||c'|| \mathrm{d}t = r(c(t_0)) + \int_{t_0}^1 ||c'|| \mathrm{d}t.$$

By the Intermediate-Value Theorem, there will be a first value t_1 for which $r(c(t_1)) = ||v||$. For this choice we get

$$L[c] = ||v|| + \int_{t_1}^1 ||c'|| \mathrm{d}t.$$

Thus $L(c) = ||v||$ if and only if wherever $c(t)$ is smooth we have $c'(t) = \lambda(t)\frac{\partial}{\partial r}$ with $\lambda(t) \geq 0$ and $||c'|| \equiv 0$ for $t > t_1$. Then we may as well assume $t_1 = 1$. Moreover, up to reparameterization, each smooth segment of c is a segment of a radial geodesic. But then, since c is continuous, it follows that c is actually, up to reparameterization, a (single smooth) radial geodesic. In particular, if $L[c] = \rho(c(0), c(1))$, then we conclude that up to reparameterization $c(t)$ is a (smooth) geodesic.

(2) Let $c(t)$ be a curve from p to q. Since $q \notin B_r(p)$, there is a first value t_0 such that $c(t_0) \in \partial B_r(p)$. By the above,

$$L[c] \geq r + \rho(c(t_0), q) \geq r + \rho(\partial B_r(p), q).$$

Therefore

$$\rho(p,q) = \inf_c L[c] \geq r + \rho(\partial B_r(p), q).$$

But by triangle inequality the opposite inequality is also true. Therefore

$$\rho(p,q) = r + \rho(\partial B_r(p), q).$$

Since $\partial B_r(p)$ is compact, there exists $q' \in \partial B_r(p)$ such that $\rho(q', q) = \rho(\partial B_r(p), q)$. This suffices to complete the proof. □

4. The Hopf-Rinow Theorem

The preceding discussion suggests two natural questions:

(1) When is $\exp_p$ defined on all of M_p?

(2) When is it possible to join two arbitrary points by a geodesic whose length is equal to the distance between them?

The answers to these questions are related and given by the Hopf-Rinow Theorem.

THEOREM 1.10. *The following are equivalent:*

(a) M is a complete metric space where the distance from p to q in M is defined as the minimum length of all curves from p to q.

(b) For some $p \in M$, $\exp_p$ is defined on all of M_p.

(c) For all $p \in M$, $\exp_p$ is defined on all of M_p.

Any of these conditions imply

(d) Any two points p, q of M can be joined by a geodesic whose length is the distance from p to q.

In practice, the implication $(a) \Rightarrow (d)$ is most important. (a) is a very natural hypothesis which holds in particular whenever M is compact. On the other hand, it is necessary to know (d) in order to apply geometrical and analytical tools to the study of M. Theorems 1.31, 1.39, 1.42, as well as various theorems in later chapters, illustrate this. We shall often use the implication $(a) \Rightarrow (d)$ without explicitly mentioning Theorem 1.10.

In the sequel we shall assume that all manifolds are complete.

PROOF OF THEOREM 1.10. $(b) \Rightarrow (a)$. We shall first show that if for some p, $\exp_p$ is defined on all M_p, then any point q can be connected to p

by a geodesic whose length is the distance from p to q. We then show that this statement together with (b) implies (a). Now, given p, let $B_r(p)$ be a normal coordinate ball. By Corollary 1.9(1) we may assume that $q \notin B_r(p)$, so by Corollary 1.9(2) let $q' \in \partial B_r(p)$ be such that

$$\rho(p, q) = r + \rho(q', q).$$

Let $\gamma : [0, \infty) \to M$ be the normal geodesic such that $\gamma|[0, r]$ is the unique minimal geodesic from p to q'. The set of t such that

$$\rho(p, \gamma(t)) + \rho(\gamma(t), q) = \rho(p, q)$$

is clearly closed, so let $t_0 \in [r, \rho(p, q)]$ be the last such value. Let $B_{r_1}(\gamma(t_0))$ be a normal coordinate ball about $\gamma(t_0)$ so that there exists $q'' \in \partial B_{r_1}(\gamma(t_0))$ such that

$$\rho(\gamma(t_0), q) = \rho(\gamma(t_0), q'') + \rho(q'', q).$$

Let σ be the unique minimal geodesic from $\gamma(t_0)$ to q''. Since

$$\rho(p, q) = \rho(p, \gamma(t_0)) + \rho(\gamma(t_0), q'') + \rho(q'', q),$$

by triangle inequality we have

$$\rho(p, \gamma(t_0)) + \rho(\gamma(t_0), q'') = \rho(p, q'').$$

But

$$L[\gamma]|[0, t_0] = \rho(p, \gamma(t_0)), \qquad L[\sigma] = \rho(\gamma(t_0), q'').$$

Therefore

$$L[\gamma \cup \sigma] = \rho(p, q'').$$

The the curves γ and σ must fit together to form a smooth geodesic $\gamma \cup \sigma = \gamma|[0, t_0 + r_1]$. Then

$$\rho(p, q) = \rho(p, \gamma(t_0 + r_1)) + \rho(\gamma(t_0 + r_1), q),$$

which is a contradiction.

To finish the proof that $(b) \Rightarrow (a)$, let q_i be a Cauchy sequence, and let $\gamma_i : [0, t_i] \to M$ be a sequence of minimal normal geodesics with $\gamma(t_i) = q_i$. Clearly $\{t_i\}$ is also a Cauchy sequence with limit say t_0. By compactness of the unit sphere at p we may pass to a subsequence such that $\gamma'_{i_j}(0) \to v$. Let $\gamma : [0, \infty) \to M$ be the normal geodesic such that $\gamma'(0) = v$. Then the theory of ordinary differential equations (continuous dependence of solutions on initial data)applied to the geodesic equation gives $q_{i_j} = \gamma(t_{i_j}) \to \gamma(t_0)$. Since q_i is a Cauchy sequence, in fact $q_i \to \gamma(t_0)$, which completes the proof.

$(a) \Rightarrow (c)$ We must show that given p and $v \in M_p$ there exists a geodesic $\gamma : [0, 1] \to M$ such that $\gamma'(0) = v$. Let $[0, t_0)$ be the largest open interval for which such a γ exists. Then if $t_i \uparrow t_0$, $\gamma(t_i)$ is a Cauchy sequence with limit say q. Define $\gamma(t_0) = q$. Then $\gamma|[0, t_0]$ is continuous. Let $B_r(p)$ be a normal coordinate ball. For i sufficiently large, $\gamma(t_i) \in B_r(q)$. Let $\sigma : (-r, r) \to M$ be the unique minimal geodesic such that $\gamma(t_i) \in \sigma$ and $\sigma(0) = q$. Then $\gamma \cup \sigma$ is continuous, piecewise smooth and

$$L\big[\gamma \cup \sigma|[t_0 - r, t_0 + r]\big] = 2r.$$

Hence $\gamma \cup \sigma$ is actually smooth, and γ extends past t_0, which is a contradiction.

$(c) \Rightarrow (b)$ is trivial.

$(c) \Rightarrow (d)$. This is the same argument as was already given in the first part of the proof that $(b) \Rightarrow (a)$. □

In Chapter 5 we will take up the question of finding conditions under which the minimizing geodesic between two points is unique.

5. The curvature tensor and Jacobi fields

From the Gauss Lemma we know that at $v \in M_p$ the deviation of dexp from being an isometry is measured by the extent to which it fails to preserve the inner product on vectors in P, the subspace of $(M_p)_v$ which is perpendicular to the direction v itself. This failure is in turn measured by the *curvature tensor*.

The curvature tensor R assigns to each $p \in M$ a trilinear map of $M_p \times M_p \times M_p \to M_p$. If x, y, z are elements of M_p, we extend them to vector fields X, Y, Z and define

$$R(x, y)z = \nabla_X \nabla_Y Z - \nabla_Y \nabla_X Z - \nabla_{[X,Y]} Z.$$

It is easy to check that $R(x, y)z$ does not depend on the extension to vector fields and that it is antisymmetric in x and y. Also, a straightforward computation shows that

$$R(x, y)z + R(y, z)x + R(z, x)y = 0.$$

This is called the Jacobi (or the first Bianchi) identity. Moreover,

$$\langle R(x, y)z, w\rangle = \langle R(z, w)x, y\rangle,$$

as is straightforward to check.

We now examine the relationship between curvature and the exponential map.

Let v and w be orthonormal vectors in M_p. Using the natural identification of the tangent space at any $x \in M_p$ with M_p itself, v and w induce(parallel) vector fields V and W on all of M_p. Consider the family ρ_s of rays in M_p defined by

$$\rho_s(t) = (V + sW)t.$$

Then $\exp_p \circ \rho_s$ is a geodesic through p with initial tangent vector $v + sw$. In order to measure the effect of dexp_p on the lengths of vectors in P, we shall compute some terms of the Taylor expansion of $||\mathrm{dexp}(tW)||^2$.

First some preliminaries. It is clear from the definition that $\mathrm{dexp}(tW)$ arises as the variation field of the 1-parameter family of geodesics $\exp_p \circ \rho_s$. Fields of this type are called *Jacobi fields* and are characterized as the solutions of a certain second-order differential equation. More precisely, let $\alpha(t, s) : [a, b] \times (-\epsilon, \epsilon) \to M$ be such that for fixed s, $\alpha(t, s)$ is a geodesic. Let $T = d\alpha\left(\frac{\partial}{\partial t}\right)$ and $V = d\alpha\left(\frac{\partial}{\partial s}\right)$.

We now determine the differential equation satisfied by $V|\alpha(t,0)$. By the discussion at the end of Section 1,

$$\nabla_T V - \nabla_V T - \mathrm{d}\alpha\left[\frac{\partial}{\partial t}, \frac{\partial}{\partial s}\right] = 0.$$

But $\left[\frac{\partial}{\partial t}, \frac{\partial}{\partial s}\right] = 0$, so $\nabla_T V = \nabla_V T$. Therefore $\nabla_T \nabla_T V = \nabla_T \nabla_V T$. Since $\nabla_T T = 0$, we may write

$$\nabla_T \nabla_T V = \nabla_T \nabla_V T = \nabla_T \nabla_V T - \nabla_V \nabla_T T.$$

Using the definition of the curvature tensor and the fact that

$$[T, V] = \mathrm{d}\alpha\left(\left[\frac{\partial}{\partial t}, \frac{\partial}{\partial s}\right]\right) = 0,$$

we get the *Jacobi equation*

$$\nabla_T \nabla_T V = R(T, V)T.$$

A vector field V, along a geodesic γ with tangent vector T satisfying this equation is called a *Jacobi field.* Let $\{E_i(t)\}$ be orthogonal and parallel along $\alpha(t, 0)$. Then the Jacobi equation may be written as the linear second-order system of ordinary differential equations

$$\begin{bmatrix} \langle J, E_1\rangle \\ \vdots \\ \langle J, E_n\rangle \end{bmatrix}'' = \langle R(T, E_i)T, E_j\rangle \begin{bmatrix} \vdots \\ \langle J, E_i\rangle \\ \vdots \end{bmatrix}.$$

From the theory of ordinary differential equations it follows that the space of solutions of this system is $2n$-dimensional and that there exists a unique solution with prescribed initial value and first derivative. This is equivalent to prescribing $J(0)$ and $J'(0) = \nabla_T J|_{t=0}$. Notice that since $\nabla_T T = 0$, we have

$$\langle J, T\rangle'' = \langle J'', T\rangle = \langle R(T, J)T, T\rangle = 0.$$

Therefore any Jacobi field J may be written uniquely as

$$J = J_0 + (at + b)T,$$

where $\langle J_0, T\rangle \equiv 0$, Finally, if J is a Jacobi field, then J comes from a variation of geodesics. In fact, let $c(s)$ be a curve such that $c'(0) = J(0)$, and let T and $J'(0)$ be extended to parallel fields along $c(s)$. Then the variation field of $\exp_{c(s)}(t(T + sJ'(0)))$ is a Jacobi field with the same initial conditions as J. Therefore it equals J by the uniqueness theorem above.

We are now ready to calculate the Taylor series for $||\mathrm{d}\exp(tW)||^2$. Set $\mathrm{d}\exp(V) = T$ and $\mathrm{d}\exp(tW) = J$. Then it is easily seen that $J'(0) = w$. Then

$$\begin{aligned}
&\langle J, J\rangle|_{t=0} = 0, \\
&\langle J, J\rangle'|_{t=0} = 2\langle J, J'\rangle|_{t=0} = 0, \\
&\langle J, J\rangle''|_{t=0} = 2\langle J', J'\rangle|_{t=0} + 2\langle J'', J\rangle|_{t=0} = 2||W||^2 = 2.
\end{aligned}$$

Note that in fact $J''|_{t=0} = R(T,J)T|_{t=0} = 0$, so

$$\langle J,J\rangle''' = 6\langle J'',J'\rangle|_{t=0} + 2\langle J''',J\rangle|_{t=0} = 0.$$

Also,

$$J''' = \nabla_T(R(T,J)T)|_{t=0} = (\nabla_T R)(T,J)T|_{t=0} + R(T,J')T|_{t=0}.$$

So

$$J'''|_{t=0} = R(T,J')T|_{t=0} = R(T,w)T|_{t=0}.$$

Then

$$\begin{aligned}\langle J,J\rangle'''' &= 8\langle J''',J'\rangle|_{t=0} + 6\langle J'',J''\rangle|_{t=0} + 2\langle J'''',J\rangle|_{t=0}\\ &= 8\langle R(T,w)T,w\rangle = -8\langle R(w,T)T,w\rangle.\end{aligned}$$

Therefore

$$||\mathrm{dexp}(tW)||^2 = t^2 - \frac{1}{3}\langle R(w,T)T,w\rangle t^4 + O(t^5). \tag{1.11}$$

We find that when compared to the rays ρ_s geodesics $\exp\circ\rho_s$ come together to the order of $(\frac{1}{3}\langle w,R(w,v)v\rangle)^{\frac{1}{2}}t^2$. So if $\langle R(w,v)v,w\rangle$ is positive, geodesics locally converge, and if it is negative, they locally diverge by comparison with rays.

(1.11) may also be given the following interpretation. Let $g_{ij}(x)$ denote the metric expressed in terms of normal coordinates. Then at the origin, up to first order the metric looks like the Euclidean metric $g_{ij} = \delta_{ij}$. The deviation comes in with the second-order terms, which are in turn measured by curvature.

Given any plane σ in M_p and two vectors v and w which span σ, we define the *sectional curvature* $K(\sigma)$ to be

$$\frac{\langle R(v,w)w,v\rangle}{||v\wedge w||^2}.$$

Here $||v\wedge w||^2$ denotes the square of the area of the parallelogram spanned by v and w. One can easily check that $K(\sigma)$ does not depend on the choice of the spanning vectors. Furthermore, the curvature tensor R is completely determined by the inner product together with the function $K: G_{2,n}(M_m)\to\mathbb{R}$, where $G_{2,n}(M_m)$ denotes the space of all 2-dimensional subspaces of M_m, the Grassmann manifold. In fact, a straightforward computation shows that

(1.12)

$$\begin{aligned}\langle R(x,y)z,w\rangle = \frac{1}{6}\{&K(x+w,y+z)||(x+w)\wedge(y+z)||^2\\ &-K(y+w,x+z)||(y+w)\wedge(x+z)||^2\\ &-K(x,y+z)||x\wedge(y+z)||^2-K(y,x+w)||y\wedge(x+w)||^2\\ &-K(z,x+w)||z\wedge(x+w)||^2-K(w,y+z)||w\wedge(y+z)||^2\\ &+K(x,y+w)||x\wedge(y+w)||^2+K(y,z+w)||y\wedge(z+w)||^2\\ &+K(z,y+w)||z\wedge(y+w)||^2+K(w,x+z)||w\wedge(x+z)||^2\\ &+K(x,z)||x\wedge z||^2+K(y,w)||y\wedge w||^2\\ &-K(x,y)||x\wedge w||^2-K(y,z)||y\wedge z||^2.\end{aligned}$$

Here $K(x,y)$ denotes the curvature of the plane spanned by x, y. If the curvatures of all plane sections are of the same sign, then this sign is a fundamental invariant. By studying in more detail its effect on the behavior of geodesics, we will derive topological and geometrical information. We will use the notation $K_M > H$ to indicate that for all plane sections at all points of M the sectional curvature is bigger than the constant H.

The condition $K \equiv 0$ is equivalent to the statement that in normal coordinates $g_{ij} \equiv \delta_{ij}$, as will be clear from the results of the next section and Section 14.

6. Conjugate points

As was shown in Section 2, in order for a curve γ to realize the distance between its endpoints, it is necessary that γ be a geodesic. However, if γ is too long, this condition is not sufficient. For example, on the unit sphere, geodesics are great circles. A geodesic of the length of more than π will not minimize. Suppose, for example, that γ starts at p. There are infinitely many geodesics σ having length π and going from p to the antipodal point q.

The path $\sigma \cup \tau$ shown in Fig. 1.2, consists of the segment of σ from p to r, and the minimal geodesic τ from r to s will have length shorter than the segment of γ from p to s. Now in this example q is a singular value of $\exp_p$. Although the above argument just used the fact that there were *two* distinct geodesics from p to q, the example suggests that there should also be a connection between failure of geodesics to minimize globally and singular values of exp. Intuitively, these are points at which distinct geodesics come together at least infinitesimally. To demonstrate this connection, we shall need an infinitesimal version of the argument. We begin with a characterization of the singularities of exp. We say that q is *conjugate* to p if q is a singular value of $\exp : M_p \to M$. The conjugacy is said to be along $\gamma = \gamma_v$ if dexp is singular at v. The *order* of a conjugate point is defined to be the

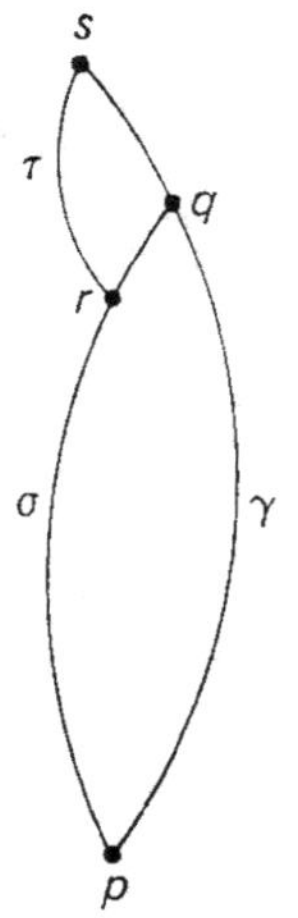

Fig. 1.2.

dimension of the null space of dexp : $(M_p)_v \to M_q$. Let us pick $v \in M_p$ and $w \in (M_p)_v$, and assume $\text{dexp}(w) = 0$. We know immediately that w (when identified with its corresponding vector in M_p) is perpendicular to v, because the length of any component of w in the v-direction is preserved by dexp. Also, if as in the previous section we form the rays $\rho_s(t) = (v + sw)t$ and geodesics $\gamma_s = \exp \circ \rho_s$, we find

$$\frac{\mathrm{d}}{\mathrm{d}s}(\exp \circ \rho_s(1)) = 0.$$

Hence there exists a one-parameter family of geodesics γ_s which come from p and whose associated Jacobi field, the variation vector field $\frac{\mathrm{d}}{\mathrm{d}s}(\gamma_s(t))$ vanishes at $q = \gamma_0(1)$. Thus we have proved half of the following proposition.

PROPOSITION 1.13. *q is conjugate to p along a geodesic γ if and only if there exists a non-zero Jacobi field J along γ such that $J(0) = J(1) = 0$. Hence q is conjugate to p if and only if p is conjugate to q.*

PROOF. Suppose there exists a nonzero J along γ with $J(0) = J(1) = 0$. $\alpha(t,s) = \exp_{\gamma(0)}(T + sJ'(0))t$ is a rectangle whose associated variation field is J. Then

$$\text{dexp}J'(0)|_{\gamma'(0)\in M_{\gamma(0)}} = J(1) = 0,$$

so p is conjugate to q. Since the Jacobi-field condition is symmetric in p and q, so is conjugacy. □

Now we give two elementary facts about Jacobi fields and conjugate points.

PROPOSITION 1.14. *Let $\gamma : [a,b] \to M$ be a geodesic, $\gamma' = T$, and assume there is a Jacobi field J which vanishes at $\gamma(a)$ and $\gamma(b)$. Then $\langle J, T\rangle \equiv \langle J', T\rangle \equiv 0$.*

PROOF. We note that

$$T\langle T, J'\rangle = \langle T, J''\rangle = -\langle T, R(J,T)T\rangle = -\langle R(T,T)J, T\rangle = 0.$$

Therefore, $\langle T, J'\rangle$ is constant on γ. But $\langle T, J'\rangle = T\langle T, J\rangle$, so that $\langle T, J\rangle$ as a function of the parameter of γ has constant derivative. But

$$\langle T, J\rangle_{\gamma(a)} = \langle T, J\rangle_{\gamma(b)} = 0,$$

so $\langle T, J\rangle \equiv 0$ and $\langle T, J'\rangle \equiv 0$. □

REMARK 1.15. This argument actually shows that

$$\langle T, J\rangle = \langle T, J(0)\rangle + \langle T, J'(0)\rangle t.$$

PROPOSITION 1.16. *If $\gamma(a)$ and $\gamma(b)$ are not conjugate, then a Jacobi field J along γ is determined by its values at $\gamma(a)$ and $\gamma(b)$.*

PROOF. Let W and J be two Jacobi fields which coincide at the endpoints of γ. Then $W - J$ is a Jacobi field which vanishes at both endpoints. But since these points are not conjugate, $W - J$ must be identically zero, so $W = J$ □

Suppose that M has constant curvature K. Then the formula of the previous section specializes to

$$\langle R(x,y)z, w\rangle = -K(\langle x, z\rangle\langle y, w\rangle - \langle y, z\rangle\langle x, w\rangle).$$

Then the Jacobi equation expressed in terms of parallel fields becomes

$$\langle J'', E_i\rangle = \langle \nabla_T\nabla_T J, E_i\rangle = -\langle R(J,T)T, E_i\rangle = K(\langle J, T\rangle\langle T, E_i\rangle - \langle J, E_i\rangle).$$

So if $\langle J, T\rangle = 0$, we get simply

$$\langle J, E_i\rangle'' = -K\langle J, E_i\rangle.$$

The reader may verify that the general solution is

$$K > 0 : \sum(a_i \sin(\sqrt{K}t) + b_i \cos(\sqrt{K}t))E_i(t);$$
$$K = 0 : \sum(a_i t + b_i)E_i(t);$$
$$K < 0 : \sum(a_i \sinh(\sqrt{-K}t) + b_i \cosh(\sqrt{-K}t))E_i(t).$$

If $K < 0$ or $K = 0$, geodesics have no conjugate points, while if $K > 0$ conjugate points occur at $t = \frac{\pi l}{\sqrt{K}}$, where l is an integer.

7. Second variation of arc length

We have shown in Section 2 that geodesics are always critical points of the arc-length function. However, as we have observed in Section 6, they are not always local minima. Therefore we compute the second derivative of arc length with respect to a variation.

Let $\gamma : [a, b] \to M$ be a geodesic, and let $\alpha : Q \to M$ be a smooth map, where Q is the rectangular solid $[a, b] \times [-\epsilon, \epsilon] \times [-\delta, \delta]$ and $\alpha(t, 0, 0) = \gamma(t)$. This means that α is a 2-parameter variation of the geodesic γ. We will look at the arc-length function differentiated successively with respect to these two parameters. Let $L(v, w)$ be the arc length of the curve $t \to \alpha(t, v, w)$, so

$$L(v, w) = \int_a^b ||T|| \mathrm{d}s.$$

Assume $||\gamma'|| \equiv 1$. Let T, V, W be vector fields corresponding to the first, second and third variables of α, respectively.

As in the proof of the first variation formula

$$\frac{\partial}{\partial v} L(v, w) = \int_a^b \frac{\langle \nabla_T V, T\rangle}{||T||}.$$

Then

$$\begin{aligned}
\frac{\partial^2}{\partial w \partial v} L(v, w) &= \frac{\partial}{\partial w} \int_a^b \frac{\langle \nabla_T V, T\rangle}{||T||} \\
&= \int_a^b \frac{\langle \nabla_W \nabla_T V, T\rangle + \langle \nabla_T V, \nabla_W T\rangle}{||T||} - \langle \nabla_T V, T\rangle \frac{\langle \nabla_W T, T\rangle}{||T||^3} \\
&= \int_a^b \frac{\langle R(W, T)V, T\rangle + \langle \nabla_T \nabla_W V, T\rangle + \langle \nabla_T V, \nabla_T W\rangle}{||T||} \\
&\quad - \frac{\langle \nabla_T V, T\rangle \langle \nabla_T W, T\rangle}{||T||^3}.
\end{aligned}$$

Using $||T|||_{(0,0)} \equiv 1$ and $\nabla_T T|_{(0,0)} \equiv 0$,

$$\begin{aligned}
\left.\frac{\partial^2 L}{\partial w \partial v}\right|_{(0,0)} &= \int_a^b \langle \nabla_T V, \nabla_T W\rangle - \langle R(W, T)T, V\rangle \\
&\quad - T\langle \nabla_W V, T\rangle - T\langle V, T\rangle T\langle W, T\rangle \\
&= -\langle \nabla_W V, T\rangle|_a^b + \int_a^b \langle \nabla_T V, \nabla_T W\rangle \\
&\quad - \int_a^b \langle R(W, T)T, V\rangle - T\langle V, T\rangle T\langle W, T\rangle.
\end{aligned} \tag{1.17}$$

This is the *second variation formula.* Note that (1.17) is valid for 2-parameter piecewise smooth variations (with the obvious definition). This follows by restricting attention to the subintervals on which V, W are smooth, then adding and observing that the endpoint terms cancel. In case the variation is through geodesics, by Proposition 1.14, $T\langle V, T\rangle$ and $T\langle W, T\rangle$ are constant. Then if $\langle V, T\rangle$ or $\langle W, T\rangle$ vanishes at both endpoints, the last term drops out. Moreover, if either V or W vanishes at the endpoints or, more generally, $\nabla_V W = 0$, we get

$$\left.\frac{\partial^2 L}{\partial w \partial v}\right|_{(0,0)} = \int_a^b \langle \nabla_T V, \nabla_T W\rangle + \langle R(W, T)V, T\rangle. \tag{1.18}$$

We remark that in this case the second variation depends only on the restrictions of V, W to γ. We call the above integral the *index form* $I(V, W)$.

It is a symmetric bilinear form on the space of piecewise smooth vector fields V, W along γ such that $\langle V, T\rangle \equiv \langle W, T\rangle \equiv 0$. Notice that I is independent of the orientation of γ. If I is positive definite on vector fields vanishing at $\gamma(a), \gamma(b)$, then γ is a minimum among all nearby curves with the same endpoints.

If $a = t_0 < t_1 < \cdots < t_n = b$ is such that $V, W|[t_i, t_{i+1}]$ is smooth, then using

$$\langle \nabla_T V, \nabla_T W\rangle = T\langle \nabla_T V, W\rangle - \langle \nabla_T \nabla_T V, W\rangle$$

and integrating, we get

$$(1.19)\quad I(V,W) = \sum_i \langle \Delta_{t_i}(\nabla_T V), W\rangle - \int_a^b \langle \nabla_T\nabla_T V, W\rangle + \langle R(V,T)T, W\rangle,$$

where

$$\Delta_{t_i}(\nabla_T V) = \lim_{t\to t_i^+} \nabla_T V - \lim_{t\to t_i^-} \nabla_T V.$$

In particular, if $V|[t_i, t_{i+1}]$ is a Jacobi field, then

$$I(V,W) = \sum_i \langle \Delta_{t_i}(\nabla_T V), W\rangle.$$

PROPOSITION 1.20. *Let I be defined on all piecewise smooth vector fields along γ which vanish at the endpoints. Then the null space of I is exactly the set of Jacobi fields along γ which vanish at $\gamma(a)$ and $\gamma(b)$. Specifically, V is a Jacobi field if and only if $I(V,W) = 0$ for all W.*

PROOF. By the above it suffices to show that $I(V,W) = 0$ for all W implies that V is a Jacobi field. Let f be a function vanishing at $\{t_i\}$ and positive elsewhere. Then setting

$$W = f(t)(-\nabla_T\nabla_T V + R(T,V)T)$$

we see that $V|[t_i, t_{i+1}]$ is a Jacobi field for all i. Now letting W_0 be a field such that $W_0(t_i) = \Delta_{t_i}\nabla_T V$, the claim follows. □

COROLLARY 1.21. *I has a non-trivial null space if and only if $\gamma(a)$ is conjugate to $\gamma(b)$ along γ. The dimension of the null space is the order of the conjugate point $\gamma(b)$.*

PROOF. The first statement is merely Proposition 1.13. The second follows from the construction used in the proof of that proposition. It provides a linear isomorphism between the null space of dexp and the space of Jacobi fields along γ which vanish at the endpoints. □

8. Submanifolds and the second fundamental form

We leave our study of the second variation of arc length in order to present some facts about connections on submanifolds which we shall need later. We return to second variation in Section 9.

Let N be a submanifold of M. Then for each $p \in N$, let $P : M_p \to N_p$ be the orthogonal projection with respect to $\langle\ ,\ \rangle$. The normal bundle $\nu(M)$

(or simply ν) is the subset of $T(M)$ defined by: $x \in \nu$ if $x \in M_p$ for $p \in N$, and $P(x) = 0$. ν is a vector bundle over N whose dimension is the difference of the dimensions of M and N. Its fiber at $p \in N$ will be denoted ν_p.

The Riemannian metric of M also induces a positive-definite bilinear form on each tangent space N_p. So N inherits a Riemannian metric, and hence an affine connection, which we shall call ∇^0. In fact ∇^0 can be defined as follows: Given X and Y vector fields on N, extend them to M and set $\nabla^0_X Y$ equal to $P(\nabla_X Y)$. It is easy to check that $\nabla^0$is the unique torsion-free connection on N which is compatible with the metric.

The *second fundamental form* S of N is the difference between ∇ and ∇^0. Specifically if x and y are in N_p, we extend them to vector fields and define $S(x, y)$ to be $\nabla_X Y - \nabla^0_X Y$. One checks that the definition is independent of the extension and that $S : N_p \times N_p \to \nu_p$ is a bilinear map. It is also symmetric, because if we extend x and y so that $[X, Y](p) \in N_p$, then

$$\nabla_X Y - \nabla_Y X = [X, Y] = \nabla^0_X Y - \nabla^0_Y X.$$

Given $z \in \nu_p$, we define $S_z : N_p \times N_p \to \mathbb{R}$ by

$$S_z(x, y) = \langle S(x, y), z\rangle.$$

Of course, S_z is a symmetric bilinear form on N_p.

Generalizing the notion of conjugate point, we say that a *focal point* q of N is a singular value of $\exp|_\nu$. We call q a focal point of N at p if there is a singular inverse image of q somewhere in ν_p.

There is a particular submanifold of codimension one which will prove useful in what follows. Fix $p \in M$ and x a unit vector in M_p. Let

$$x^\perp = \{y \in M_p | \langle x, y\rangle = 0\}.$$

Since $\exp : M_p \to M$ is a local diffeomorphism at zero, there is a neighborhood U of zero in $x^\perp$ such that $\exp|_U$ is an embedding. Let N be the submanifold $\exp(U)$. We call N the *geodesic submanifold* defined by x.

LEMMA 1.22. *$S_x = 0$, where S is the second fundamental form of N.*

PROOF. Pick $z \in N_p$. Let γ be a geodesic from p in direction z. Then near p, $\gamma \subseteq N$, so we can extend z to a vector field Z on N such that Z is the tangent vector to γ near p. Then $(\nabla_Z Z)(p) = 0$, so $S_x(z, z) = 0$. Therefore, since S_x is symmetric, $S_x = 0$. □

COROLLARY 1.23. *Let x be extended to X, the unit normal field to N, $\langle X, X\rangle \equiv 1$, $\langle X, Y\rangle \equiv 0$ for all $Y \in T(N)$. Then $\nabla_z X = 0$ for $z \in N_p$.*

PROOF. If Y is any vector field on N, then

$$\langle \nabla_Z Y, X\rangle_p = Z\langle Y, X\rangle_p - \langle Y, \nabla_Z X\rangle_p.$$

But $S_x = 0$ implies $\langle \nabla_Z Y, X\rangle_p = 0$, and $\langle Y, X\rangle$ is zero on N, so $z\langle Y, Z\rangle = 0$. Therefore $\langle y, \nabla_Z X\rangle_p = 0$ for all $y \in N_p$. But since X has constant length,

$$\langle \nabla_Z X, X\rangle = \frac{1}{2} Z\langle X, X\rangle = 0;$$

hence $\nabla_Z X = 0$. □

If S_X vanishes identically, then N is called *totally geodesic*. Let γ be a geodesic of such an N. The equation

$$\nabla_{\gamma'}\gamma' = \nabla^0_{\gamma'}\gamma' + S(\gamma', \gamma') = 0$$

shows that γ is a geodesic of M. By uniqueness of geodesics, this is equivalent to the statement that if γ is a geodesic of M tangent to N at $\gamma(0)$, then γ remains in N. Conversely, if N has the latter property, then as in Lemma 1.22, $S \equiv 0$.

9. Basic index lemmas

In this section we prove two lemmas which show that in a certain sense, Jacobi fields minimize the index form. The first lemma involves conjugate points, while the second is an analogous statement with conjugate points replaced by focal points of the submanifold N described above. These lemmas are of fundamental importance and are used in the proofs of Rauch Comparison Theorems and the Morse Index Theorem in Chapter 4.

LEMMA 1.24 (First Lemma). *Let γ be a geodesic in M from p to q such that there are no points conjugate to p on γ. Let W be a piecewise smooth vector field on γ and V the unique Jacobi field such that $V(p) = W(p) = 0$ and $V(q) = W(q)$. Then $I(V,V) \le I(W,W)$, and equality holds only if $V{=}W$.*

LEMMA 1.25 (Second Lemma). *Let γ be a geodesic in M from p to q. Let x be the tangent vector to γ at p and let N be the geodesic submanifold defined by x. Assume that N has no focal points along γ. Let W be a piecewise smooth vector field along γ, V the unique Jacobi field such that $(\nabla_X V)(p) = 0$ and $V(q) = W(q)$. Then $I(V,V) \le I(W,W)$, and equality holds only if $V = W$.*

PROOF OF FIRST LEMMA. Let $\{V_i\}$ be a basis of T_qM and extend each V_i to a Jacobi field along γ such that $V_i(p) = 0$. This is uniquely possible since γ has no conjugate points, and the V_i's are linearly independent except at p. Since $V_i(p) = 0$, we can write $V_i = tA_i$, where t is the parameter of $\gamma : [0,1] \to M$ and A_i is some vector field on γ. Then $V_i'(p) = A_i$, so $\{A_i\}$ is also linearly independent, and thus there are functions $q_i(t)$ such that $W = \sum\limits_i q_i(t)A_i$. But since $W_p = 0$, there exist piecewise smooth functions f_i so that $W = \sum f_i V_i$. Then $V = \sum f_i(1)V_i$.

We shall make two preliminary calculations:

$$I(V,V) = \langle V'(1), V(1)\rangle = \sum f_i(1)f_j(1)\langle V_i'(1), V_j(1)\rangle. \tag{1.26}$$

In fact, since V is a Jacobi field, $V'' = R(T,V)T$, so

$$\begin{aligned} I(V,V) &= \int_0^1 \langle V',V'\rangle + \langle R(T,V)T,V\rangle \\ &= \int_0^1 \langle V',V\rangle' - \langle V'',V\rangle + \langle R(T,V)T,V\rangle = \langle V'(1),V(1)\rangle. \end{aligned}$$

If V_i, V_j are Jacobi fields, then

$$\langle V_i',V_j\rangle - \langle V_i,V_j'\rangle = c, \tag{1.27}$$

for some constant c. We compute

$$\begin{aligned} (\langle V_i',V_j\rangle - \langle V_i,V_j'\rangle)' &= \langle V_i'',V_j\rangle + \langle V_i',V_j'\rangle - \langle V_i',V_j'\rangle - \langle V_i,V_j''\rangle \\ &= \langle V_i'',V_j\rangle - \langle V_i,V_j''\rangle \\ &= \langle R(T,V_i)T,V_j\rangle - \langle V_i R(T,V_j)T\rangle = 0. \end{aligned}$$

The last step follows by the usual symmetry property of the curvature tensor. In our case the constant in (1.9.2) is zero since the expression vanishes at $t=0$. Now

$$\begin{aligned} \nabla_T W &= \sum f_i'V_i + f_iV_i' = A + B, \\ I(W,W) &= \int \langle A,A\rangle + \langle A,B\rangle + \langle B,A\rangle + \langle B,B\rangle + \langle R(T,W)T,W\rangle, \\ \int \langle B,B\rangle &= \sum \int f_if_j\langle V_i'V_j'\rangle = \sum \int f_if_j(\langle V_i',V_j\rangle' - \langle V_i'',V_j\rangle). \end{aligned}$$

Integrating the first term by parts and applying the Jacobi equation to the second gives

$$\begin{aligned} \int \langle B,B\rangle = \sum & f_i(1)f_j(1)\langle V_i'(1),V_j(1)\rangle \\ & - \int (f_i'f_j\langle V_i',V_j\rangle + f_if_j'\langle V_i',V_j\rangle + \langle R(T,W)T,W\rangle). \end{aligned}$$

By (1.26), the first term is $I(V,V)$. By (1.27), the second term is $\int\langle A,B\rangle$, while the third term equals $\int\langle B,A\rangle$. Hence

$$\int \langle B,B\rangle = I(V,V) - \int \langle A,B\rangle + \langle B,A\rangle + \langle R(T,W)T,W\rangle.$$

Therefore, referring to the original expression for $I(W,W)$,

$$I(W,W) = I(V,V) + \int \langle A,A\rangle.$$

The second term on the right is nonnegative and vanishes only if $W = V$. □

We emphasize that the above calculation, in particular the integration by parts, depends only on W being piecewise smooth.

To prove the second lemma we pick $\{v_i\}$ an orthonormal basis of T_pM such that $v_1 = T$, and extend its members to Jacobi fields $\{V_i\}$ such that $(\nabla_T V_i)_p = 0$. We need the following:

SUBLEMMA 1.28. *If* $\{V_i\}$ *are as above and* N, γ *are as in the second lemma, then* $\{V_i\}$ *are linearly independent along* γ *if and only if* N *has no focal point.*

PROOF. $V_1 = T$ all along γ and $V_2, \ldots, V_n$ are everywhere perpendicular to T. Hence it suffices to show that $\{V_2, \ldots, V_n\}$ are independent or to show that if Z is any nonzero Jacobi field on γ such that $\langle Z, T\rangle_p = 0$ and $(\nabla_T Z)_p = 0$, then Z is never zero.

Let η be the geodesic from p in direction Z, and extend T along η so that $||T||$ is constant and T is perpendicular to N. Construct a geodesic from each point of η in direction T. This one-parameter family of geodesics gives rise to a Jacobi field Z_0 on γ. Clearly $Z_0(p) = Z(p)$. Also $(\nabla_{Z_0}T)(p) = 0$ by Corollary 1.23, so $(\nabla_T Z_0)(p) = 0$ and $Z_0 = Z$. From the construction of Z_0 it is immediate that Z_0 has a zero implies $\exp \nu$ has a singularity over γ. Also, if there is such a singularity, one constructs Z_0 in a straightforward manner. This proves the sublemma. □

Now the proof of the second lemma proceeds exactly as that of the first. $\langle \nabla_T V_i, V_i\rangle_p$, $\langle B, W\rangle_p$ and $\langle \nabla_T V, V\rangle_p$ are all zero because $\nabla_T V_i$ and $\nabla_T V$ are zero at p. □

We can now state that geodesics minimize locally up to the first conjugate point among curves with the same end points. More precisely, if $\gamma(t_0)$ is the first conjugate point of $\gamma(0)$ along γ, then for vector fields vanishing at $\gamma(0), \gamma(t)$ with $t < t_0$, the second variation is positive; that is, the index form is positive definite. This follows by taking $W(q) = 0$ in Lemma (1.24). Moreover, we have the following important converse.

COROLLARY 1.29. *Let* $\gamma : [0, \infty) \to M$ *be a geodesic, and let* $\gamma(t_0)$ *be conjugate to* $\gamma(0)$. *Then* $\gamma|[0,t]$ *is not minimal for* $t > t_0$.

PROOF. We can assume that $\gamma(t_0)$ is the first point conjugate to $\gamma(0)$. Let J be a nonzero Jacobi field along $\gamma[0, t_0]$ such that $J(0) = J(t_0) = 0$. Extend J to a vector field X on all of γ by declaring $X(t) = 0$ for $t > t_0$. Clearly $I(X, X) = 0$ on $[0, t]$, but X is not smooth at t_0.

Fix δ small enough so that there are no conjugate pairs on $\gamma|[t_0 - \delta, t_0 + \delta]$, and define a vector field V by:

$V = J$ on $[0, t_0 - \delta]$,

$V =$ Jacobi field W on $[t_0 - \delta, t_0 + \delta]$ such that

$$W(t_0 - \delta) = J(t_0 - \delta), \qquad W(t_0 + \delta) = 0,$$

$V = 0$ on $[t_0 + \delta, t]$ (See Fig. 1.3)

Since X is not smooth on $[t_0 - \delta, t_0 + \delta]$, it is definitely not a Jacobi field. Hence on $[t_0 - \delta, t_0 + \delta]$ we have $I(V, V) < I(X, X)$. Since $X = V$ outside this interval, in fact on $[0, t]$ we have $I(V, V) < I(X, X) = 0$. Since any V arises from a variation, it follows that there is a variation which keeps the endpoints fixed and decreases the length of γ.

□

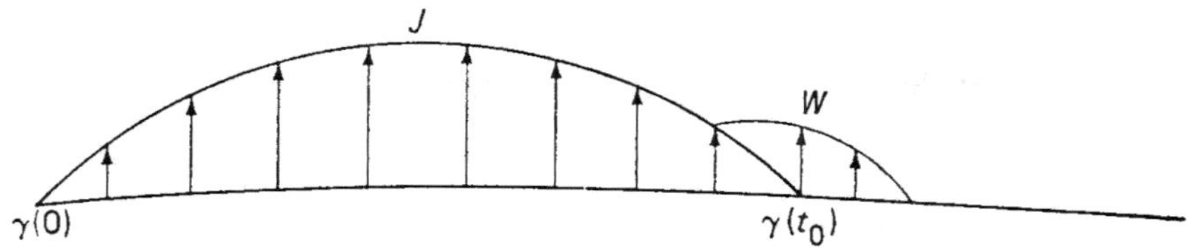

Fig. 1.3.

10. Ricci curvature and Myers' and Bonnet's Theorems

As an application of the First Index Lemma (1.22), we now prove the theorems of Myers and Bonnet. We define the *diameter* $d(M)$ of M to be the supremum of $\rho(p, q)$ for $p, q \in M$.

DEFINITION 1.30. The *Ricci curvature* is a symmetric bilinear form on each M_p of $T(M)$ defined to be the trace of the linear tranformation $z \to R(z, x)y$. Hence

$$\mathrm{Ric}(x, y) = \sum_i \langle R(e_i, x)y, e_i\rangle,$$

where $\{e_i\}$ is an orthonormal basis of M_p.

THEOREM 1.31 (Myers and Bonnet). *Let M^n be a complete Riemannian manifold. If*

(1) (Myers) *for all unit vectors x, $\mathrm{Ric}(x, x) \geq (n-1)H$, or*

(2) (Bonnet) $K_M \geq H$,

then every geodesic of length $\geq \pi/\sqrt{H}$ has conjugate points. Hence the diameter of M satisfies $d(M) \leq \pi/\sqrt{H}$.

PROOF. Fix a normal geodesic $\gamma : [0, l] \to M$, and let $\{E_i\}$ be an orthonormal basis of parallel fields along γ such that $E_n = \gamma' = T$. Let $W_i = \sin(\frac{\pi t}{l})E_i(t)$ be vector fields on γ. Then

$$\begin{aligned} I(W_i, W_i) &= -\int_0^l \langle W_i, \nabla_T^2 W_i + R(W_i, T)T\rangle \mathrm{d}t \\ &= \int_0^l (\sin(\frac{\pi t}{l}))^2 (\frac{\pi^2}{l^2} - \langle R(E_i, T)T, E_i\rangle)\mathrm{d}t. \end{aligned}$$

Thus if, for any i, $\langle R(E_i, T)T, E_i\rangle \geq H$ and $l \geq \pi/\sqrt{H}$, then $I(W_i, W_i) \leq 0$. Also,

$$\sum_{i=1}^{n-1} I(W_i, W_i) = \int_0^l (\sin(\frac{\pi t}{l}))^2((n-1)\frac{\pi^2}{l^2} - \mathrm{Ric}(T, T))\mathrm{d}t.$$

So if $\mathrm{Ric}(T, T) \geq (n-1)H$ and $l \geq \frac{\pi}{\sqrt{H}}$, then the sum and therefore at least one summand must be nonpositive. But if γ had no conjugate points, the First Index Lemma 1.24 would imply that there is a Jacobi field J such that $I(J, J) < 0$ and J would vanish at $\gamma(0)$ and $\gamma(l)$. This is impossible, so γ

has conjugate points. By Corollary 1.29, γ is not minimal. More directly, since some $I(W_i, W_i)$ is negative, the second variation of arc length is in some direction negative, so γ cannot be minimal. The theorem follows. □

The following is our first result which illustrates the influence of the sign of the curvature on the topology of M.

COROLLARY 1.32. *Let M be complete. If there exists a constant $H > 0$ such that for all unit vectors x we have* $\mathrm{Ric}(x, x) > (n-1)H > 0$, *then M is compact and has finite fundamental group.*

To prove this corollary, we use the notion of local isometry.

Let M and N be Riemannian manifolds with metrics $\langle\langle\ ,\ \rangle\rangle$ and $\langle\ ,\ \rangle$, and let $\phi : M \to N$ be a smooth map. We say ϕ is a *local isometry* if for all $p \in M$ and $v, w \in M_p$,

$$\langle\langle v, w\rangle\rangle = \langle \mathrm{d}\phi(v), \mathrm{d}\phi(w)\rangle.$$

ϕ is an *isometry* if ϕ is local isometry and also a diffeomorphism.

This notion of isometry is equivalent to the usual notion of isometry of M and N as metric spaces. The implication in one direction is easy, but it is not so obvious that every metric space isometry is smooth. This is an essential step in the Theorem of Myers and Steenrod quoted in Chapter 3. It is easy to check that a local isometry respects Riemannian connections and maps geodesics to geodesics.

Note also that if ϕ is a local isometry, then $\mathrm{d}\phi$ must be everywhere nonsingular; for if $\mathrm{d}\phi(v) = 0$, then

$$\langle v, v\rangle = \langle \mathrm{d}\phi(v), \mathrm{d}\phi(v)\rangle = 0,$$

so v is the zero vector. Conversely, if $d\phi$ is everywhere nonsingular, then we can use the Riemannian metric of N to induce one on M. For $v, w \in M_p$, we define $\langle\langle v, w\rangle\rangle$ to be $\langle \mathrm{d}\phi(v), \mathrm{d}\phi(w)\rangle$. It is easy to check that $\langle\langle\ ,\ \rangle\rangle$ is a Riemannian metric and that ϕ is a local isometry with respect to the metrics $\langle\langle\ ,\ \rangle\rangle$ on M and $\langle\ ,\ \rangle$ on N.

PROOF OF COROLLARY. Let M_c be the universal covering space of M. Since $\pi : M_c \to M$ is a local diffeomorphism, it induces a Riemannian structure on M_c, and the curvature tensor R_c at $p_c \in M_c$ is isomorphic to R at $\pi(p_c) \in M$. Therefore by Myers' theorem, $d(M) \leq \pi H^{-\frac{1}{2}}$, so M_c is compact. Hence the set $\pi^{-1}(p)$ must have finite cardinality and the corollary follows. □

11. Rauch Comparison Theorem

We will often study a Riemannian manifold when our only given infirmation consists of bounds on its sectional curvature. The following theorems allow us to draw geometrical conclusions from such information by comparing lengths in M to corresponding lengths in a manifold M_0 whose curvature is suitably related to that of M. We usually take M_0 to have constant curvature, in which case the lengths in M_0 can be calculated explicitly.

THEOREM 1.33 (First Theorem (Rauch)). *Let M^n, M_0^{n+k}, be Riemannian manifolds ($\dim M_0 \geq \dim M$), and let $\gamma, \gamma_0 : [0, l] \to M, M_0$ be normal geodesics, and set $\gamma' = T$, $\gamma_0' = T_0$. Assume that for each $t \in [0, l]$ and any $X \in M_{\gamma(t)}$, $X_0 \in (M_0)_{\gamma_0(t)}$, the sectional curvatures of the sections σ, σ_0 spanned by X, T and X_0, T_0 satisfy $K(\sigma_0) \geq K(\sigma)$. Assume further that for no $t \in [0, l]$ is $\gamma_0(t)$ conjugate to $\gamma_0(0)$ along γ_0. Let V, V_0 be Jacobi fields along γ, γ_0 such that $V(0), V_0(0)$ are tangent to γ, γ_0,*

$$||V(0)|| = ||V_0(0)||, \quad \langle T, V'(0)\rangle = \langle T_0, V_0'(0)\rangle, \quad \textit{and} \quad ||V'(0)|| = ||V_0'(0)||.$$

Then for all $t \in [0, l]$,

$$||V(t)|| \geq ||V_0(t)||.$$

THEOREM 1.34 (Second Theorem (Berger)). *Let the notation be as above, and assume that for all $t \in [0, l]$ and plane sections σ, σ_0 as above, $K(\sigma_0) \geq K(\sigma)$. Assume further that for no $t \in [0, l]$ is $\gamma(t)$ a focal point of the geodesic submanifold N_0 defined by T_0. Let V, V_0 be Jacobi fields along γ, γ_0 satisfying $V'(0), V_0'(0)$ are tangent to γ, γ_0, and $||V'(0)|| = ||V_0'(0)||$, $\langle T, V(0)\rangle = \langle T_0, V_0(0)\rangle$, $||V(0)|| = ||V_0(0)||$. Then for all $t \in [0, l]$,*

$$||V(t)|| \geq ||V_0(t)||.$$

The theorems will be referred to in the sequel as Rauch I and Rauch II. Before proving them, we will investigate a few of their consequences.

Notice first of all that by continuity it is only necessary to assume that $[0, l)$ (rather than $[0, l]$) is free of conjugate (respectively focal) points. Also, suppose that the hypothesis $K(\sigma_0) \geq K(\sigma)$ holds for all $t \in [0, \infty)$. It then follows that the first conjugate (respectively focal) point along γ occurs no sooner than the first conjugate (respectively focal) point along γ_0. In fact, if γ_0 has no conjugate point on $[0, l)$ then by the theorem, for any Jacobi field as above, $||V(t)|| \geq ||V_0(t)||$. But $||V_0(t)|| > 0$ by assumption, so $||V(t)|| > 0$. *In particular, if the sectional curvature of M satisfies $K_M \leq K$, and $B_r(0)$ denotes the ball of radius r in the tangent space at $m \in M$, then $\exp_m|B_r(0)$ is nonsingular for $r < \pi/\sqrt{K}$.*

COROLLARY 1.35 (Corollary of Rauch I). *Let M, M_0 be Riemannian manifolds with $\dim M_0 \geq \dim M$, and let $m, m_0 \in M, M_0$. Assume $K_{M_0} \geq K_M$, i.e., for all plane sections $\sigma, \sigma_0 \in M, M_0$, $K(\sigma_0) \geq K(\sigma)$. Let r be chosen such that $\exp_m|B_r(0)$ is an imbedding and $\exp_{m_0}|B_r(0)$ is nonsingular. Let $I : M_m \to M_{m_0}$ be a linear injection preserving inner products. Then for any curve $c : [0, 1] \to \exp_m(B_r(0))$, we have*

$$L[c] \geq L[\exp_{m_0} \circ I \circ \exp_m^{-1}(c)] = L[c_0(t)].$$

PROOF OF COROLLARY. Let $\tilde{c} : [0, 1] \to B_r(0)$ be the unique curve in $B_r(0)$ such that $\exp_m \tilde{c}(s) = c(s)$. Consider the rectangle $\alpha(t, s) \to$

$\exp_m(t\tilde{c}(s))$. For fixed s, the associated variation field V_s is a Jacobi field along the geodesic $\gamma_s = t \to \exp_m(t\tilde{c}(s))$ with $V_s(1) = c'(s)$. Then

$$V_s = d\exp_m(t\tilde{c}'(s)) = td\exp_m(\tilde{c}'(s)).$$

Therefore $\nabla_T V_s = \tilde{c}'(s)$. Similarly, associated to the rectangle $\alpha_0(t,s) \to \exp_{m_0} \circ I(t\tilde{c}(s))$ there is a Jacobi field V_{0s} with

$$V_{0s}(1) = c_0'(s), \qquad \nabla_T V_{0s} = (I \circ (\tilde{c}'(s)))' = I \circ (\tilde{c}'(s)).$$

Since I preserves lengths,

$$||c_0'(s)|| = ||I(c'(s))||.$$

By Rauch I,

$$||c'(s)|| = ||V_s(1)|| \geq ||V_{0s}(1)|| = ||c_0'(s)||,$$

and by integrating this inequality we are done. □

COROLLARY 1.36 (Corollary of Rauch II). *Let γ, γ_0 be geodesics on M and M_0 parameterized on $[0,l]$, with tangent vectors T and T_0. Let E and E_0 be parallel unit vectors along γ and γ_0 which are everywhere perpendicular to T and T_0. Let $c : [0,l] \to M$ be a smooth curve defined by*

$$c(t) = \exp(f(t)E(t)),$$

where $f : [0,l] \to \mathbb{R}$ is a smooth function, and let $c_0 : [0,l] \to M_0$ be defined by

$$c_0(t) = \exp(f(t)E_0(t)).$$

Assume that $K_{M_0} \geq K_M$, and assume that for each t the geodesic $\eta_0 : [0,l] \to M_0$ defined by

$$\eta_0(s) = \exp(sf(t)E_0(t))$$

contains no focal points of the geodesic submanifold defined by $\eta_0'(0)$. Then

$$L[c] \geq L[c_0].$$

PROOF. Since c and c_0 are both parameterized from 0 to l, it suffices to compare the lengths of their tangent vectors.

Fix $t_1 \in [0,l]$. Let η be the geodesic

$$\eta(s) = \exp(sf(t_1)E(t))$$

and let

$$h(t) = \exp(f(t_1)E(t)), \qquad h_0(t) = \exp(f(t_1)E_0(t)).$$

Then

$$c'(t_1) = h'(t_1) + f'(t_1)\eta'(1),$$

while

$$c_0'(t_1) = h_0'(t_1) + f_0'(t_1)\eta_0'(1).$$

By the Gauss Lemma, these sums decompose $c'(t_1)$ and $c_0'(t_1)$ into pairs of perpendicular vectors. Since $E(t_1)$ and $E_0(t_1)$ are both unit vectors,

$$||f'(t_1)\eta'(1)|| = ||f'(t_1)\eta_0'(1)||.$$

Therefore we need only compare h' and h'_0. but h' and h'_0 are tangents to the families of geodesics $\tau_t(s) = \exp(sE(t))$ and $\tau_{0t}(s) = \exp(sE_0(t))$. Therefore h' and h'_0 can be extended to Jacobi fields V and V_0 along η and η_0, and since E and E_0 are parallel, these fields satisfy the hypotheses of the Second Rauch Theorem, that is

$$\nabla_{\eta'} V = \nabla_T E = 0, \qquad \nabla_{\eta'_0} V_0 = \nabla_{T_0} E_0 = 0.$$

The corollary follows.[1] □

Proof of the First Rauch Theorem 1.33. First assume that V, V_0 are perpendicular to T, T_0 or that

$$||V(0)|| = \langle T, V'(0)\rangle = 0, \qquad ||V_0(0)|| = \langle T_0, V'_0(0)\rangle = 0.$$

Consider the ratio $\frac{||V||^2}{||V_0||^2}$, as a function of the parameter t of the geodesics. Since $V_0 = 0$ only at $\gamma_0(0)$, this is well defined everywhere except at $t = 0$. By L'Hôpital's rule (taking two derivatives),

$$\lim_{t\to 0} \frac{||V||^2}{||V_0||^2} = \frac{\langle V', V'\rangle}{\langle V'_0, V'_0\rangle} = 1.$$

Therefore to show that $||V|| \geq ||V_0||$ it suffices to show that

$$\frac{\mathrm{d}}{\mathrm{d}t}\left(\frac{||V||^2}{||V_0||^2}\right) \geq 0.$$

Equivalently, for $t > 0$,

$$\frac{\langle V', V\rangle}{\langle V, V\rangle} \geq \frac{\langle V'_0, V_0\rangle}{\langle V_0, V_0\rangle}.$$

Fix $t_1 \in [0, l)$ and define vector fields

$$W_{t_1} = \frac{V(t)}{||V(t_1)||} \qquad W_{0t_1} = \frac{V_0(t)}{||V_0(t_1)||}.$$

Then $||W(t_1)|| = ||W_0(t_1)|| = 1$, and since W_{t_1} is a constant multiple of V,

$$\frac{\langle V, V'\rangle}{\langle V, V\rangle} = \frac{\langle W_{t_1}, W'_{t_1}\rangle}{\langle W_{t_1}, W_{t_1}\rangle}, \qquad \frac{\langle V_0, V'_0\rangle}{\langle V_0, V_0\rangle} = \frac{\langle W_{0t_1}, W'_{0t_1}\rangle}{\langle W_{0t_1}, W_{0t_1}\rangle}.$$

In particular,

$$\left.\frac{\langle V, V'\rangle}{\langle V, V\rangle}\right|_{t_1} = \langle W'_{t_1}, W_{t_1}\rangle|_{t_1}, \qquad \left.\frac{\langle V_0, V'_0\rangle}{\langle V_0, V_0\rangle}\right|_{t_1} = \langle W'_{0t_1}, W_{0t_1}\rangle|_{t_1}.$$

[1]One might conjecture that Corollary 1.36 could be strengthened so as to compare lengths of any two curves which have the same expression in Fermi coordinates (see Hicks [1965], p. 133, for definition). Somewhat surprisingly this turns out not to be possible.

Then

$$\begin{aligned}\frac{\langle V, V'\rangle}{\langle V, V\rangle}\bigg|_{t_1} &= \langle W'_{t_1}, W_{t_1}\rangle|_{t_1}\\ &= \int_0^{t_1} \langle W'_{t_1}, W_{t_1}\rangle'\\ &= \int_0^{t_1} \langle W'_{t_1}, W'_{t_1}\rangle + \langle W''_{t_1}, W_{t_1}\rangle\\ &= \int_0^{t_1} \langle W'_{t_1}, W'_{t_1}\rangle - \langle R(W_{t_1}, T)T, W_{t_1}\rangle\\ &= \int_0^{t_1} \langle W'_{t_1}, W'_{t_1}\rangle - K(\sigma)||W_{t_1}||^2,\end{aligned}$$

where σ is the plane section spanned by T and W_{t_1}.

Let $P_{-\gamma}$ denote parallel translation along γ with the opposite parameterization and P_{γ_0} parallel translation along γ_0. Let $I : M_{\gamma(0)} \to (M_0)_{\gamma_0(0)}$ be an inner product-preserving injection and define $I_t : M_{\gamma(t)} \to (M_0)_{\gamma_0(t)}$ by

$$I_t(X) = P_{\gamma_0} \circ I \circ P_{-\gamma}(X).$$

Assume that I has been chosen so that $I_t(T) = T_0$ and $I_{t_1}(W_{t_1}) = W_{0t_1}$. Define a field $\widehat{W}_{0t_1}$ by

$$\widehat{W}_{0t_1} = I_t(W_{t_1}(t)).$$

Since $\widehat{W}_{0t_1}$ has the same expression in terms of parallel frames as W_{t_1}, clearly

$$\begin{aligned}\langle W_{t_1}(t), W_{t_1}(t)\rangle &= \langle \widehat{W}_{0t_1}(t), \widehat{W}_{0t_1}(t)\rangle,\\ \langle W'_{t_1}(t), W_{t_1}(t)\rangle &= \langle \widehat{W}'_{0t_1}(t), \widehat{W}_{0t_1}(t)\rangle.\end{aligned}$$

Using this and our assumption on the curvatures gives

$$\begin{aligned}\int_0^{t_1} &\langle W'_{t_1}, W'_{t_1}\rangle - \langle R(W_{t_1}, T)T, W_{t_1}\rangle\\ &\geq \int_0^{t_1} \langle \widehat{W}'_{0t_1}, \widehat{W}'_{0t_1}\rangle - \langle R_0(\widehat{W}_{0t_1}, T_0)T_0, \widehat{W}_{0t_1}\rangle\\ &\geq \int_0^{t_1} \langle W'_{0t_1}, W'_{0t_1}\rangle - \langle R_0(W_{0t_1}, T)T, W_{0t_1}\rangle,\end{aligned}$$

where the last inequality is just the First Index Lemma 1.24. But reversing the first part of our argument shows that this last expression is equal to

$$\frac{\langle V_0, V_0'\rangle|_{t_1}}{\langle V_0, V_0\rangle}.$$

Thus for arbitrary t_1,

$$\frac{\langle V', V\rangle}{\langle V, V\rangle}\bigg|_{t_1} \geq \frac{\langle V_0', V_0\rangle}{\langle V_0, V_0\rangle}\bigg|_{t_1}.$$

and we are done.

In the general case let

$$V = \widehat{V} + \langle T, V\rangle T, \qquad V_0 = \widehat{V_0} + \langle T_0, V_0\rangle T_0.$$

Then $||\widehat{V}(t)|| \geq ||\widehat{V_0}(t)||$ as above. Also,

$$\langle T, V\rangle = \langle T, V(0)\rangle + \langle T, V'(0)\rangle t = \langle T_0, V_0\rangle,$$

so $||V(t)|| \geq ||V_0(t)||$. □

Note that the equality of the index form and the term $\langle W'_{t_0}, W_{t_0}\rangle$ may be seen geometrically as follows. Let $\tilde{c}(s)$ be a curve in $M_{\gamma(0)}$ such that $||\tilde{c}(s)|| = 1$ and

$$\mathrm{dexp}(t_1 \tilde{c}'(s))|_{s=0} = W_{t_1}(t_1).$$

Consider the rectangle $\alpha : (t, s) \to \exp_{\gamma(0)}(t\tilde{c}(s))$. Then $\mathrm{d}\alpha(\frac{\partial}{\partial s})$ is a vector field along α extending W_{t_1}. We will denote this field by W_{t_1} also. Then

$$\begin{aligned}\langle W'_{t_1}, W_{t_1}\rangle &= \langle \nabla_{W_{t_1}} T, W_{t_1}\rangle = W_{t_1}\langle T, W_{t_1}\rangle - \langle T, \nabla_{W_{t_1}} W_{t_1}\rangle \\ &= -\langle T, \nabla_{W_{t_1}} W_{t_1}\rangle\end{aligned}$$

The curves $t \to \exp_{\gamma(0)}(t\tilde{c}(s))$ are all geodesics of length t_1. Therefore the second variation formula for the variation α gives

$$\langle \nabla_{W_{t_1}} W_{t_1}, T\rangle + I(W_{t_1}, W_{t_1}) = 0,$$

where I is the index form. This, together with the above, yields

$$\langle W'_{t_1}, W_{t_1}\rangle = I(W_{t_1}, W_{t_1}).$$

Proof of the Second Rauch Theorem 1.34. In this case $||V||_0 = ||V_0||_0$ and V_0 is never zero since γ_0 contains no focal point. Therefore we need only show

$$\frac{\mathrm{d}}{\mathrm{d}t}\Big(\frac{||V||^2}{||V_0||^2}\Big) \geq 0.$$

This follows in the same way as in the previous proof by the use of the Second Index Lemma 1.25. □

Remark 1.37. It is interesting to study the case $||V(t)|| = ||V_0(t)||$ in Theorems 1.33 and 1.34. One finds immediately that all the inequalities of the proofs must be equalities. Also, by lemma 1.24,

$$\widehat{W_{0l}} = W_{0l}.$$

Thus $\langle R(V, T)T, V\rangle = \langle R(V_0, T_0)T_0, V_0\rangle$.

Warner [1966] gives technical generalizations of the Rauch Theorems.

12. The Cartan-Hadamard Theorem

An easy application of Rauch I gives us some information about manifolds of nonpositive sectional curvature. We begin with a lemma.

LEMMA 1.38. *If $\phi : M^n \to N^n$ is a local isometry and M is complete, then ϕ is a covering map; i.e. for all $p \in N$ there exists a neighborhood U of p such that $\phi^{-1}(U) = \bigcup_\alpha U_\alpha$, a union of disjoint subsets of M, and for each α, $\phi|U_\alpha : U_\alpha \to U$ is a diffeomorphism.*

PROOF. Fix $p \in N$ and assume r is small enough so that a ball about p of radius r is properly contained in a normal coordinate neighborhood of p. Let U be the ball of radius r about p. Let $\{q_\alpha\} = \phi^{-1}(p)$, and let U_α be the ball of radius r about q_α. We will show that $\phi^{-1}(U)$ is the disjoint union $\bigcup_\alpha U_\alpha$, and that $\phi : U_\alpha \to U$ is a diffeomorphism for each α.

First let $B_r(0), B_r^\alpha(0)$ be the balls about zero of radius r in N_p, M_{q_α}, respectively. Since ϕ is a local isometry, the diagram

$$\begin{array}{ccc} M_{q_\alpha} & \xrightarrow{\mathrm{d}\phi} & N_p \\ \downarrow{\scriptstyle \exp_{q_\alpha}} & & \downarrow{\scriptstyle \exp_p} \\ M & \longrightarrow & N \end{array}$$

commutes. It restricts to the diagram

$$\begin{array}{ccc} B_r^\alpha(0) & \longrightarrow & B_r(0) \\ \downarrow & & \downarrow \\ U_\alpha & \longrightarrow & U \end{array}$$

and since $\exp_p \circ \mathrm{d}\phi : B_r^\alpha(0) \to U_\alpha$ is a diffeomorphism, so is $\phi : U_\alpha \to U$.

It is clear that $\cup U_\alpha \subseteq \phi^{-1}(U)$. We shall show the opposite inclusion. Given $\bar{q} \in \phi^{-1}(U)$, let $q = \phi(\bar{q})$, and let γ be the normal minimal geodesic from q to p. Let $v = \mathrm{d}\phi^{-1}(\gamma'(q))$, and let $\bar{\gamma}$ be the geodesic from $\bar{q}$ in direction v. Since M is complete, $\bar{\gamma}$ may be extended arbitrary far. Let $t_0 = \rho(p, q)$ and set $q = \bar{\gamma}(t_0)$.

Since $\phi \circ \bar{\gamma} = \gamma$, $\phi(\bar{p}) = p$. Also $\rho(\bar{p}, \bar{q}) = \rho(p, q) < r$. Hence $q \in \bigcup_\alpha U_\alpha$.

It remains to show that $U_\alpha \cap U_\beta$ is empty if $\alpha \neq \beta$. For this it is clearly sufficient to show that if $\bar{p}_\alpha, \bar{p}_\beta \in \phi^{-1}(p)$, then $\rho(\bar{p}_\alpha, \bar{p}_\beta) > 2r$. Let γ be a minimal geodesic from $\bar{p}_\alpha$ to $\bar{p}_\beta$. Then $\gamma = \phi \circ \bar{\gamma}$ is a closed geodesic on p. Therefore, since p is contained in a normal coordinate neighborhood of radius $> r$, γ must have length more than $2r$. Therefore $\rho(\bar{p}_\alpha, \bar{p}_\beta) > 2r$. The lemma follows. □

THEOREM 1.39 (Cartan-Hadamard). *Let M be complete and $K_M \leq 0$. Then for any $p \in M$, $\exp_p : M_p \to M$ is a covering map. Hence the universal covering space of M is diffeomorphic to $\mathbb{R}^n$. Hence the homotopy groups $\pi_i(M)$ vanish for $i > 1$.*

PROOF. Using Rauch I and comparing with Euclidean space, we find that M has no conjugate points. Hence $\exp_p : M_p \to M$ has nonsingular differential. Therefore we can let $\exp_p$ induce a metric $\langle\langle\ ,\ \rangle\rangle$ on M_p which makes it a local isometry. The lines through the origin of M_p are geodesics in this metric because they are mapped by $\exp_p$ into geodesics on M. Hence by the Hopf-Rinow Theorem 1.10, M_p is complete in the metric $\langle\langle\ ,\ \rangle\rangle$. Therefore by Lemma 1.38, $\exp_p$ is a covering map. It is a standard fact from the theory of covering spaces that if $\bar{M}$ covers M,

$$\pi_i(\bar{M}) = \pi_i(M) \qquad \text{for } i \geq 2.$$

Hence

$$\pi_i(M) = \pi_i(\mathbb{R}^n) = 0.$$

□

COROLLARY 1.40. *Let M^n be complete, simply connected and have non-positive curvature. Then M is diffeomorphic to $\mathbb{R}^n$.*

PROOF. A covering map onto a simply connected space must be a homeomorphism. $\exp_p$ is smooth and nonsingular, so it is a diffeomorphism. □

13. The Cartan-Ambrose-Hicks Theorem

We shall prove a theorem which tells how under suitable conditions relating the curvatures of M^n and $\bar{M}^n$, we can construct an isometry between them.

We begin with a local result. Fix $p \in M^n$, $\bar{p} \in \bar{M}^n$ and let $I : M_p^n \to \bar{M}_{\bar{p}}^n$ be a linear isometry. Let $B_r(p)$ be a normal coordinate neighborhood of p. Define $\phi : B_r(p) \to B_r(\bar{p})$ by $\phi = \exp_{\bar{p}} \circ I \circ \exp_p^{-1}$. Then if r is sufficiently small, $B_r(\bar{p})$ is a normal coordinate about $\bar{p}$ and ϕ is a diffeomorphism. Let P_γ denote parallel translation along a geodesic γ. Let $R, \bar{R}$ denote the curvature tensors in $M, \bar{M}$ and $\bar{\gamma} = \phi(\gamma)$. Set $I_\gamma = P_{\bar{\gamma}} \circ I \circ P_{-\gamma}$. The following lemma is the local version of the fact that behavior of the curvature tensor under parallel translation determines the metric.

LEMMA 1.41. *In the above situation suppose that for all geodesics γ emanating from p we have*

$$I_\gamma(R(x,y)z) = \bar{R}(I_\gamma(x), I_\gamma(y))I_\gamma(z),$$

Then ϕ is an isometry and $\mathrm{d}\phi = I_\gamma$.

PROOF. Given $x \in M_q$, let γ be the geodesic from p to $q = \gamma(t^*)$ lying in $B_r(p)$ and let J be the Jacobi field along γ such that $J(0) = 0$ and $J(t^*) = x$. Let $\gamma_t = \gamma|[0,t]$, and define $\bar{J}$ along $\bar{\gamma}$ by $\bar{J}(t) = I_{\gamma_t}(J(t))$. It follows immediately from the hypothesis that $\bar{J}(t)$ is a Jacobi field along $\bar{\gamma}$. Moreover, clearly

$$||J(t)|| = ||\bar{J}(t)||.$$

To complete the proof it will suffice to show that $\bar{J}(t) = \mathrm{d}\phi(J(t))$. From the relation

$$\bar{J}(t) = P_{\gamma_t} \circ I \circ P_{-\gamma_t}(J(t))$$

it follows that $I(J'(0)) = \bar{J}'(0)$. Since $J, \bar{J}$ are Jacobi fields vanishing at $t = 0$, we have, as in Section 6.

$$J(t) = \mathrm{dexp}_{\gamma(0)} tJ'(0)|_{t\gamma'(0)}, \qquad \bar{J}(t) = \mathrm{dexp}_{\bar{\gamma}(0)} t\bar{J}'(0)|_{t\bar{\gamma}'(0)}.$$

Then

$$\begin{aligned}\bar{J}(t) &= \mathrm{dexp}_{\bar{\gamma}(0)} I(t\bar{J}'(0))|_{t\bar{\gamma}'(0)} \\ &= \mathrm{dexp}_{\bar{\gamma}(0)} \circ \mathrm{d}I \circ \mathrm{dexp}^{-1}_{\gamma(0)}(J(t)) = \mathrm{d}\phi(J(t)),\end{aligned}$$

which completes the proof. □

Now let M be complete. We proceed to a global version of the ablove lemma.

A broken geodesic is a continuous curve $\gamma : [0, l] \to M$ such that there exists $0 < t_0 < t_1 < \cdots < t_n < l$ and $\gamma|[t_i, t_{i+1}]$ is a smooth geodesic. Set

$$i\gamma = \gamma|[0, t_i],$$

and define v_i by

$$\gamma|[t_i, t_{i+1}] = t \to \exp_{\gamma(t_i)}((t - t_i)v_i).$$

If $I : M_p \to \bar{M}_{\bar{p}}$, we define a correspondence between broken geodesics emanating from p, $\bar{p}$ as follows: Set

$$\bar{\gamma}_1(t) = \exp_{\gamma(0)}(tI(v_0)).$$

Assume $\bar{\gamma}_i$ is already defined.

Set

$$\bar{\gamma}_{i+1}(t) = \begin{cases} \bar{\gamma}_i(t), & 0 \leq t \leq t_i \\ \exp_{\bar{\gamma}_i(t_i)}(t(P_{\bar{\gamma}_i} \circ I \circ P_{-\bar{\gamma}_i}(v_i))), & t_i \leq t \leq t_{i+1}. \end{cases}$$

Note that this is consistent with the definition of $\bar{\gamma}$ preceeding Lemma 1.41.

THEOREM 1.42 (Cartan, Ambrose, Hicks). *Let M^n, $\bar{M}^n$ be complete, M^n simply connected and $I : M^n_p \to \bar{M}^n_{\bar{p}}$ be a linear isometry. Suppose that for all broken geodesics γ,*

$$I_\gamma(R(x, y)z) = \bar{R}(I_\gamma(x), I_\gamma(y))I_\gamma(z).$$

Then for all broken geodesics γ_0, γ_1 from p such that $\gamma_0(l_0) = \gamma_1(l_1)$ we have

$$\bar{\gamma}_0(l_0) = \bar{\gamma}_1(l_1).$$

Thus there is a map $\Phi : M^n \to \bar{M}^n$ defined by $\gamma(l) \to \bar{\gamma}(\bar{l})$. Moreover, Φ is a local isometry and hence a covering map.

PROOF. (1) First assume that $\gamma_0(l_0) = \gamma_1(l_1)$, and that $\gamma_0, \gamma_1, \bar{\gamma}_0, \bar{\gamma}_1$ are contained in normal coordinate balls $B_r(p), B_r(\bar{p})$, respectively. Then Lemma 1.41 implies that the map $\phi = \exp_{\bar{p}} \circ I \circ \exp_p^{-1} | B_r(p)$ is an isometry. It follows that $\phi(\gamma_i) = \bar{\gamma}_i$. Therefore $\bar{\gamma}_0(l_0) = \bar{\gamma}_1(l_1)$ and $d\phi = I_{\gamma_0} = I_{\gamma_1}$.

(2) Now, for convenience and without real loss of generality, we will suppose that γ_0, γ_1 both have n breaks at the points $t_1 < \cdots < t_n$, and $l_0 = l_1 = l$. Assume that for all i, $\gamma_1(t_{i+2}), \gamma_1(t_{i+1}), \gamma_0(t_{i+1})$ and the minimal segments between them lie in a normal coordinate ball about $\gamma_0(t_i)$ and that the same is true for $\bar{\gamma}_1(t_{i+2}), \bar{\gamma}_1(t_{i+1}), \bar{\gamma}_0(t_{i+1}), \bar{\gamma}_0(t_i)$. Let $\tau : (t_{n-1}, t_n] \to M$ be the minimal geodesic from $\gamma_0(t_{n-1})$ to $\gamma_1(t_n)$. By induction we may assume that

$$\overline{{}_{n-1}\gamma_0 \cup \tau(t_n)} = \overline{{}_n\gamma_1(t_n)}$$

and that

$$I_{n-1\gamma_0 \cup \tau} = I_{{}_n\gamma_1}. \qquad (*)$$

Also the isometry $I_{n-1\gamma_0} : M_{\gamma_0(t_{n-1})} \to \bar{M}_{\bar{\gamma}_0(t_{n-1})}$ induces a correpondence (which we denote with a double bar) between geodesics emanating from $\gamma_0(t_{n-1})$ and $\bar{\gamma}_0(t_{n-1})$. Set

$$\sigma_0 = \gamma_0 | [t_{n-1}, l], \qquad \theta_1 = \gamma 1 | [t_n, l].$$

Applying step (1) gives

$$\overline{\overline{\sigma_0}}(l) = \overline{\overline{\tau \cup \theta_1}}(l), \qquad I_{\sigma_0} = I_{\tau \cup \theta_1}.$$

But this is equivalent to

$$\bar{\gamma}_0(l) = \overline{{}_{n-1}\gamma_0 \cup \tau \cup \theta_1}(l), \qquad I_{\gamma_0} = I_{{}_{n-1}\gamma_0 \cup \tau \cup \theta_1}.$$

Using $(*)$, the isometry on the right can be rewritten as

$$P_{\bar{\theta}_1} \circ I_{{}_{n-1}\gamma_0 \cup \tau} \circ P_{-\theta_1} = P_{\bar{\theta}_1} \circ I_{{}_n\gamma_1} \circ P_{-\theta_1} = I_{\gamma_1}.$$

In particular, $\bar{\gamma}_0(l) = \bar{\gamma}_1(l)$.

(3) Now let γ_0 and γ_1 be any two broken geodesics such that $\gamma_0(l) = \gamma_1(l)$. Since M is simply connected, there is a homotopy h_s from γ_0 to γ_1. By uniform continuity of h_s we may choose subdivisions $0 < s_1 < \cdots < s_m < 1$ and $0 < t_1 < \cdots < t_n < l$ such that for all i, j, the points $h_{s_{j+1}}(t_{i+2})$, $h_{s_{j+1}}(t_{i+1})$, $h_{s_j}(t_{i+1})$ and the geodesics between them lie in a normal coordinate neighborhood of $h_{s_j}(t_i)$. By inserting further breakpoints we can assume that $\{t_i\}$ is exactly the set of breakpoints of γ_0, γ_1. Let γ_{s_j} denote the broken geodesic formed by minimal segments from $h_{s_j}(0)$ to $h_{s_j}(t_1)$, $h_{s_j}(t_1)$ to $h_{s_j}(t_2)$ .... Since from the theory of ordinary differential equations the correspondence $\gamma \to \bar{\gamma}$ is continuous in an obvious sense, we may assume that for all j, $\bar{\gamma}_{s_j}$ and $\bar{\gamma}_{s_{j+1}}$ are sufficiently close, so that for all i, $\bar{\gamma}_{s_{j+1}}(t_{i+2}), \bar{\gamma}_{s_{j+1}}(t_{i+1}), \bar{\gamma}_{s_j}(t_{i+1})$ lie in a normal coordinate neighborhood of $\bar{\gamma}_{s_j}(t_i)$. Therefore, each pair $\bar{\gamma}_{s_j}, \bar{\gamma}_{s_{j+1}}$ satisfies the hypothesis of step (2). Therefore

$$\bar{\gamma}_0(l) = \bar{\gamma}_{s_1}(l) = \cdots = \bar{\gamma}_1(l).$$

Finally, let $q \in M$ be arbitrary, and let γ be a geodesic such that $\gamma(l) = q$. Then by Lemma 1.41 the map

$$\phi = \exp_{\bar{\gamma}(l)} \circ I_\gamma \circ \exp^{-1}_{\gamma(l)}$$

is an isometry from a neighborhood of $B_r(\gamma(l))$ to $B_r(\bar{\gamma}(l))$. But referring to the definition of the correspondence $\gamma \to \bar{\gamma}$ and of Φ, one sees that $\phi = \Phi | B_r(q)$. Therefore Φ is a local isometry. □

14. Spaces of constant curvature

The simplest examples of Riemannian manifolds are those whose sectional curvature is a constant K. The complete ones are called *space forms*. We will show that for each K all simply connected spaces forms with curvature K are isometric. They may be described as follows:

(a) $K = 0$. Let $M^n = \mathbb{R}^n$ with the usual metric.

(b) $K > 0$. Let $M^n = S^n_{\frac{1}{\sqrt{K}}}$, the sphere in $\mathbb{R}^{n+1}$, with the induced metric.[2]

(c) $K < 0$. Let M^n be the open set in $\mathbb{R}^n$ defined by

$$M = \left\{ x \in \mathbb{R}^n \middle| \; ||x||^2 < -\frac{4}{K} \right\}.$$

Using the standard coordinates in $\mathbb{R}^n$, define the metric by

$$\langle v, w \rangle = \frac{\sum_{i=1}^n v_i w_i}{1 + \frac{1}{4} K \sum_{i=1}^n (x_i)^2},$$

where $v, w \in M_x$.

Recall from Section 6 that if M has constant curvature K, then

$$R(x, y)z = K(\langle z, y \rangle x - \langle z, x \rangle y).$$

THEOREM 1.43. *Let M^n and $\bar{M}^n$ be complete simply connected manifolds with constant curvature K. Then M^n and $\bar{M}^n$ are isometric.*

In fact, given any $p \in M^n$, $\bar{p} \in \bar{M}^n$ and an isometry $I : M^n_p \to \bar{M}^n_{\bar{p}}$, there exists an isometry $\Phi : M^n \to \bar{M}^n$ such that $\Phi(p) = \bar{p}$ and $d\Phi_p = I$.

PROOF. This is immediate from the Cartan-Ambrose-Hicks Theorem 1.42 and the formula above for $R(x, y)z$. □

Theorem 1.42 (or Lemma 1.41) shows in particular that the vanishing of the sectional curvature is a necessary and sufficient condition for M to be locally isometric to Euclidean space.

[2] As a matter of notation, we will denote by S^n_r the n-sphere of radius r. We denote by S^n the unit sphere and by S^n_p the tangent space at $p \in S^n$.

CHAPTER 2

Toponogov's Theorem

In this chapter we shall prove Toponogov's Theorem, which is a powerful global generalization of the first Rauch Theorem. There will be two equivalent statements, and it will be convenient to prove them simultaneously.

All indices below are to be taken modulo 3.

DEFINITION 2.1. A *geodesic triangle* in the Riemannian manifold M is a set of three geodesic segments parameterized by arc length $(\gamma_1, \gamma_2, \gamma_3)$ of lengths l_1, l_2, l_3 such that $\gamma_i(l_i) = \gamma_{i+1}(0)$ and $l_i + l_{i+1} \geq l_{i+2}$. Set

$$\alpha_i = \sphericalangle(-\gamma'_{i+1}(l_{i+1}), \gamma'_{i+2}(0)),$$

the angle between $-\gamma'_{i+1}(l_{i+1})$ and $\gamma'_{i+2}(0)$, $0 \leq \alpha_i \leq \pi$.

We shall specify a geodesic triangle by giving its sides $(\gamma_1, \gamma_2, \gamma_3)$.

THEOREM 2.2 (Toponogov). *Let M be a complete manifold with $K_M \geq H$.*

(A) Let $(\gamma_1, \gamma_2, \gamma_3)$ determine a geodesic triangle in M. Suppose γ_1, γ_3 are minimal and if $H > 0$, suppose $L[\gamma_2] \leq \frac{\pi}{\sqrt{H}}$. Then in M^H, the simply connected 2-dimensional space of constant curvature H, there exists a geodesic triangle $(\bar{\gamma}_1, \bar{\gamma}_2, \bar{\gamma}_3)$ such that $L[\gamma_i] = L[\bar{\gamma}_i]$ and $\bar{\alpha}_1 \leq \alpha_1$, $\bar{\alpha}_3 \leq \alpha_3$. Except in case $H > 0$ and $L[\gamma_i] = \frac{\pi}{\sqrt{H}}$ for some i, the triangle in M^H is uniquely determined.

(B) Let γ_1, γ_2 be geodesic segments in M such that $\gamma_1(l_1) = \gamma_2(0)$ and $\sphericalangle(-\gamma'_1(l_1), \gamma'_2(0)) = \alpha$. We call such a configuration a hinge l and denote it by $(\gamma_1, \gamma_2, \alpha)$. Let γ_1 be minimal, and if $H > 0$,

$$L[\gamma_2] \leq \frac{\pi}{\sqrt{H}}.$$

Let $\bar{\gamma}_1, \bar{\gamma}_2 \subset M^H$ be such that $\gamma_1(l_1) = \gamma_2(0)$, $L[\gamma_i] = L[\bar{\gamma}_i] = l_i$ and $\sphericalangle(-\bar{\gamma}'_1(l_1), \bar{\gamma}'_2(0)) = \alpha$. Then

$$\rho(\gamma_1(0), \gamma_2(l_2)) \leq \rho(\bar{\gamma}_1(0), \bar{\gamma}_2(l_2)).$$

Proof. The proof consists of a number of steps, none of which is difficult. In what follows, conditions involving $\sqrt{H}$ should be ignored if $H \leq 0$. At first we work with $M^{H-\epsilon}$ instead of M^H. This is really crucial only for steps (5)-(7), and even then only if $H > 0$.

(1) Let $\bar{\gamma}_1, \bar{\gamma}_2 \subset M^{H-\epsilon}$, $\bar{\gamma}_1(l_1) = \bar{\gamma}_2(0)$, $\sphericalangle(\bar{\gamma}_1'(l_1), \bar{\gamma}_2'(0)) = \alpha$, and $L[\gamma_i] \leq \frac{\pi}{\sqrt{H}}$. As α increases from 0 to π, $\rho(\bar{\gamma}_1(0), \bar{\gamma}_2(l_2)) = f(\alpha)$ increases strictly monotonically from $|l_1 - l_2|$ to $D = \min\{\frac{2\pi}{\sqrt{H-\epsilon-l_1-l_2}}, l_1 + l_2\}$.

PROOF. Think of $\bar{\gamma}_1$ as held fixed and $\bar{\gamma}_2$ as varying. If $H \leq 0$, it is clear that $\rho(\bar{\gamma}_1(0), \bar{\gamma}_2(l_2))$ is a smooth function of α (see Theorem 1.39). This is also true for $0 \leq \alpha \leq \pi$ and $H > 0$. To see this, it suffices to show that $\rho(\bar{\gamma}_1(0), \bar{\gamma}_2(l_2)) < \frac{\pi}{H-\epsilon}$. If

$$\rho(\bar{\gamma}_1(0), \bar{\gamma}_2(l_2)) = \frac{\pi}{H-\epsilon},$$

then, $\bar{\gamma}_1 \cup \bar{\gamma}_2$ is a broken geodesic between antipodal points on the 2-sphere of curvature $H - \epsilon$. Such a geodesic is always smooth if the length of each segment is less than $\frac{\pi}{\sqrt{H-\epsilon}}$. Therefore, $\rho(\bar{\gamma}_1(0), \bar{\gamma}_2(l_2)) = \frac{\pi}{\sqrt{H-\epsilon}}$ implies $\alpha = \pi$.

As $\bar{\gamma}_2(l_2)$ moves, it traces out a circle of radius l_2 about $\bar{\gamma}_2(0)$. The minimal geodesic $\bar{\sigma}$ from $\bar{\gamma}_1(0)$ to $\bar{\gamma}_2(l_2)$ is perpendicular to this circle only for $\alpha = 0$ or π. Otherwise $\sigma \cup -\bar{\gamma}_2$ would form a smooth geodesic from $\bar{\gamma}_1(0)$ to $\bar{\gamma}_1(l_1)$, distinct from γ_1. This is impossible for $H \leq 0$ and possible only if $l_1 = \frac{\pi}{\sqrt{H-\epsilon}}$ for $H > 0$, which is ruled out by assumption. (See Fig. 2.1).

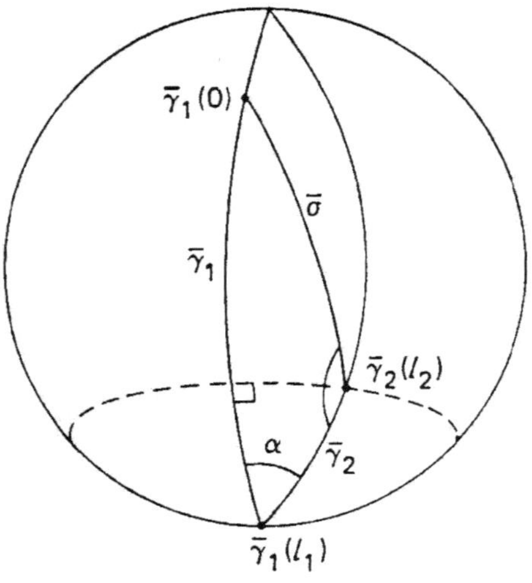

Fig. 2.1.

Hence by the first variation formula (1.5), $f'(\alpha) \neq 0$ for $\alpha \in (0, \pi)$, so f is strictly monotone. Clearly $f(0) = |l_2 - l_1|$ and $f(\pi) = D$. Since $D > |l_2 - l_1|$, f must be increasing. □

(2) In $M^{H-\epsilon}$ a triangle with sides of length $\leq \frac{\pi}{\sqrt{H}}$ is determined up to congruence by the lengths of its sides.

PROOF. Let $\{\bar{\sigma}_i\}$ determine a triangle in $M^{H-\epsilon}$ with $L[\bar{\sigma}_i] = L[\bar{\gamma}_i]$.

For fixed $L[\bar{\gamma}_1], L[\bar{\gamma}_2]$, it follows from (1) that $L[\bar{\gamma}_3]$ uniquely determines α_3. Then, by homogeneity of $M^{H-\epsilon}$ (Theorem 1.43), there exists an isometry taking $\bar{\sigma}_1$ onto $\bar{\gamma}_1$, $\bar{\sigma}_2$ onto $\bar{\gamma}_2$ and hence $\bar{\sigma}_3$ onto $\bar{\gamma}_3$. □

(3) Fix a hinge $(\gamma_1, \gamma_2, \gamma_3)$ such that γ_1 is minimal and $L[\gamma_2] \leq \frac{\pi}{\sqrt{H}}$. Then the following statements (A') and (B') are equivalent.

(A') Let γ_3 be any minimal geodesic from $\gamma_2(l_2)$ to $\gamma_1(0)$. Then there exists a triangle $(\bar{\gamma}_1, \bar{\gamma}_2, \bar{\gamma}_3)$ in $M^{H-\epsilon}$ with $L[\gamma_i] = L[\bar{\gamma}_i]$ and $\bar{\alpha}_3 \leq \alpha_3$.

(B') Let $(\bar{\gamma}_1, \bar{\gamma}_2, \alpha_3)$ be a hinge in $M^{H-\epsilon}$ with $L[\gamma_i] = L[\bar{\gamma}_i]$, $i = 1, 2$. Then

$$l_2 - l_1 \leq \rho(\gamma_1(0), \gamma_2(l_2)) \leq \rho(\bar{\gamma}_1(0), \bar{\gamma}_2(l_2)).$$

PROOF. $(A') \Rightarrow (B')$. Given $(\gamma_1, \gamma_2, \alpha_3)$ as in (B'), form the geodesic triangle $(\gamma_1, \gamma_2, \gamma_3)$, where γ_3 is a minimal geodesic from $\gamma_1(0)$ to $\gamma_2(l_2)$. We may assume that $\{\gamma_1, \gamma_2, \gamma_3\}$ form a triangle, i.e.

$$l_1 + l_3 \geq l_2, \qquad l_2 + l_3 \geq l_1.$$

Otherwise the implication is trivial. By (A'), there is a triangle $(\bar{\gamma}_1, \bar{\gamma}_2, \bar{\gamma}_3)$ in $M^{H-\epsilon}$ with $L[\gamma_i] = L[\bar{\gamma}_i]$ and $\bar{\alpha}_3 \leq \alpha_3$. Then as $\bar{\alpha}_3$ is increased to α_3 holding $L[\bar{\gamma}_1]$ and $L[\bar{\gamma}_2]$ constant, $\rho(\bar{\gamma}_1(0), \bar{\gamma}_2(l_2))$ is nondecreasing by (1). Note that this part of the argument did not use minimality of γ_1 or γ_2.

$(B') \Rightarrow (A')$. Given the situation of (A'), by (B'), if $\bar{\gamma}_1, \bar{\gamma}_2 \subset M^{H-\epsilon}$ and $\sphericalangle(-\bar{\gamma}_1'(l_1), \bar{\gamma}_2'(0)) = \alpha_3$, then

$$\rho(\bar{\gamma}_1(0), \bar{\gamma}_2(l_2)) \geq \rho(\gamma_1(0), \gamma_2(l_2)).$$

By (1) and the fact that $(\gamma_1, \gamma_2, \gamma_3)$ satisfies the triangle inequality, if the angle between $\bar{\gamma}_1, \bar{\gamma}_2$ is decreased sufficiently to an angle $\bar{\alpha}_3$, we will then have

$$\rho(\bar{\gamma}_1(0), \bar{\gamma}_2(l_2)) = \rho(\gamma_1(0), \gamma_2(l_2)).$$

Now if $\bar{\gamma}_3$ is the minimal geodesic from $\bar{\gamma}_1(l_1)$ to $\bar{\gamma}_2(l_2)$, $\bar{\gamma}_1, \bar{\gamma}_2, \bar{\gamma}_3$ is a geodesic triangle with $L[\bar{\gamma}_i] = L[\gamma_i]$ and $\bar{\alpha}_3 \leq \alpha_3$. □

Notice that we have used minimality of γ_3 (and γ_1) not for the existence of the triangle in $M^{H-\epsilon}$ but for the angle comparison.

Let $(\gamma_1, \gamma_2, \alpha)$ determine a hinge. We call this hinge *small* if $\frac{1}{2}r = \max L[\gamma_i]$, $i = 1, 2$ and $\exp_{\gamma_2(0)}|B_r(0)$ is an embedding. Let $(\gamma_1, \gamma_2, \gamma_3)$ determine a triangle. Call this triangle *small* if each of the hinges $(\gamma_i, \gamma_{i+1}, \alpha_{i+2})$ is small.

(4) (A) holds for small triangles and (B) holds for small hinges.

PROOF. We show directly that (B) holds for small hinges, and apply (3).

If $(\gamma_1, \gamma_2, \alpha_3)$ is a hinge, let $(\bar{\gamma}_1, \bar{\gamma}_2, \alpha)$ be a hinge in $M^{H-\epsilon}$ with $L[\bar{\gamma}_i] = L[\gamma_i]$, $i = 1, 2$. Let $\gamma_1(l_1) = p$, and $\bar{\gamma}_1(l_1) = \bar{p}$. Let $\bar{\gamma}_3$ be a minimal geodesic from $\bar{\gamma}_1(0)$ to $\bar{\gamma}_2(l_2)$ and $I : M_{\bar{p}}^{H-\epsilon} \to M_p$ be an injective isometry such that

$$I(\bar{\gamma}_1'(l_1)) = \gamma_1'(l_1), \qquad I(\bar{\gamma}_2'(0)) = \gamma_2'(0).$$

Let c be a curve in M defined by

$$c = \exp_p \circ I \circ \exp_{\bar{p}}^{-1}(\bar{\gamma}_3);$$

then c connects $\gamma_1(0)$ and $\gamma_2(l_2)$, and since the hinge is small, we may apply Corollary 1.35 of Rauch I to get $L[c] \leq L[\bar{\gamma}_3]$. Therefore (B) holds for small hinges.

Fix a vertex, say $\gamma_2(0)$, of the triangle $(\gamma_1, \gamma_2, \gamma_3)$. By (3), there exists a triangle in $M^{H-\epsilon}$ with the same side lengths as our given triangle and $\bar{\alpha}_3 \leq \alpha_3$. By (2) a triangle in $M^{H-\epsilon}$ is determined by the lengths of its sides. Therefore, if we start by fixing some other vertex, we will obtain the same triangle in $M^{H-\epsilon}$. Therefore there is a unique triangle in $M^{H-\epsilon}$ with $L[\bar{\gamma}_i] = L[\gamma_i]$ and $\bar{\alpha}_i \leq \alpha_i$, which proves (A) for small triangles.

Let $(\gamma_1, \gamma_2, \frac{1}{2}\pi)$ determine a hinge. Let $(\bar{\gamma}_1, \bar{\gamma}_2, \frac{1}{2}\pi)$ determine a hinge in $M^{H-\epsilon}$ with $L[\bar{\gamma}_i] = L[\gamma_i]$. Let $\bar{\gamma}_3$ be the minimal segment from $\bar{\gamma}_2(l_2)$ to $\bar{\gamma}_1(0)$. We can write

$$\bar{\gamma}_3(t) = \exp_{\bar{\gamma}_2(t)} f(t).\bar{E}(t),$$

where $\bar{E}(t)$ is the unit parallel field along γ_2 perpendicular to γ_2' and $f(t)$ is the appropriate function. (If $\gamma_2(t)$ is parameterized by arc length, γ_3 will not be, but this is unimportant). Let $E(t)$ be the parallel field along γ_2 with $E(0) = -\gamma_1'(l_1)$. We call $(\gamma_1, \gamma_2, \frac{1}{2}\pi)$ a *thin right hinge* if the hypothesis of Corollary 1.36 of Rauch II applies to the curves $\exp f(t)E(t)$ and $\exp_{\bar{\gamma}_2(t)} f(t)\bar{E}(t) = \bar{\gamma}_3(t)$. This is equivalent to demanding that for no t is there a focal point along the geodesic $s \to \exp_{\gamma_2(t)} sE(t)$ for $s < f(t)$. ☐

(5) (B) holds for thin right hinges.

PROOF. This is immediate from Corollary 1.36 to the Second Rauch Theorem. ☐

Let $(\gamma_1, \gamma_2, \alpha)$ determine a hinge with $\alpha > \frac{1}{2}\pi$. Let $(\bar{\gamma}_1, \bar{\gamma}_2, \alpha)$ be the corresponding hinge in $M^{H-\epsilon}$ with $L[\bar{\gamma}_i] = L[\gamma_i]$. Let $\bar{\gamma}_3$ denote the minimal geodesic from $\bar{\gamma}_2(l_2)$ to $\bar{\gamma}_1(0)$.

Let $\bar{\sigma} : [0, l] \to M^{H-\epsilon}$ denote the geodesic segment emanating from $\gamma_2(0)$ such that $\langle \bar{\sigma}'(0), \bar{\gamma}_2'(0) \rangle = 0$, $\bar{\sigma}'(0)$ is of the form

$$-\delta\gamma_1'(l_1) + \beta\gamma_2'(0),$$

with $\delta, \beta > 0$, and $\bar{\sigma}(l)$ is the first point of σ which lies on $\bar{\gamma}_3$. (See Fig. 2.2).

Let σ denote the geodesic segment emanating from $\gamma_2(0)$ such that

$$\langle \sigma'(0), \gamma_2'(0) \rangle = 0, \qquad \sigma'(0) = -\delta\gamma_1'(0) + \beta\gamma_2'(0).$$

with $\delta, \beta > 0$ and $L[\sigma] = L[\bar{\sigma}] = l$ Define $(\gamma_1, \gamma_2, \alpha)$ to be a *thin obtuse hinge* if $(\gamma_1, \sigma, \alpha - \frac{1}{2}\pi)$ is a small hinge and $(\sigma, \gamma_2, \frac{1}{2}\pi)$ is a thin right hinge.

(6) (B) holds for thin obtuse hinges.

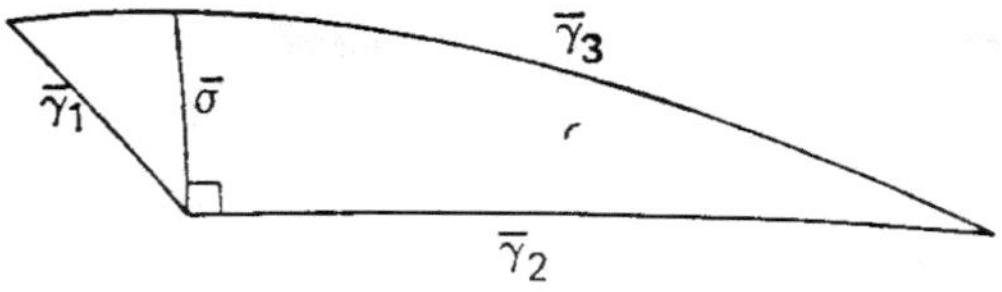

Fig. 2.2.

PROOF. By the triangle inequality,

$$\rho(\gamma_1(0), \gamma_2(l_2)) \le \rho(\gamma_1(0), \sigma(l)) + \rho(\sigma(l), \gamma_2(l_2)).$$

By (4) and (5),

$$\rho(\gamma_1(0), \sigma(l)) \le \rho(\bar{\gamma}_1(0), \bar{\sigma}(l)), \qquad \rho(\sigma(l), \gamma_2(l_2)) \le \rho(\bar{\sigma}(l), \bar{\gamma}_2(l_2)).$$

Then

$$\rho(\gamma_1(0), \gamma_2(l_2)) \le \rho(\bar{\gamma}_1(0), \bar{\sigma}(l)) + \rho(\bar{\sigma}(l), \bar{\gamma}_2(l_2)) = \rho(\bar{\gamma}_1(0), \bar{\gamma}_2(l_2)).$$

□

Let $(\gamma_1, \gamma_2, \alpha)$ determine a hinge with $\alpha < \frac{1}{2}\pi$. Let $\gamma_2(l)$ be a point on γ_2 closest to $\gamma_1(0)$, let

$$\tau = \gamma_2|[0, l], \qquad \theta = \gamma_2|[l, l_2]$$

and $\sigma : [0, k] \to M$ a minimal geodesic from $\gamma_1(0)$ to $\gamma_2(l)$. We call $(\gamma_1, \gamma_2, \alpha)$ a *thin acute hinge* if (γ_1, τ, σ) is a small triangle, $0 < l < l_2$ and $(\sigma, \theta, \frac{1}{2}\pi)$ is a thin right hinge. (See Fig. 2.3)

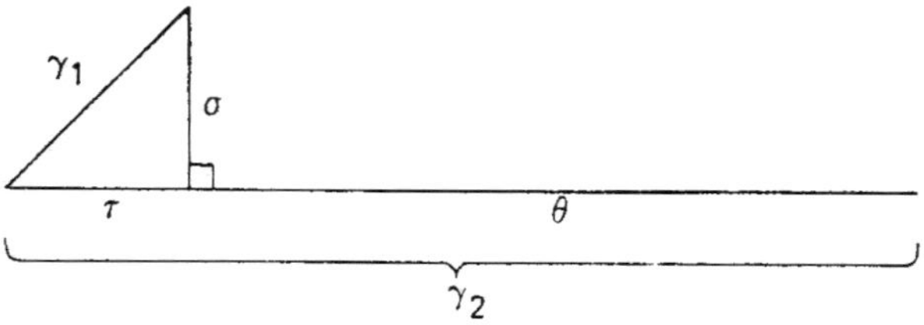

Fig. 2.3.

(7) (B) holds for thin acute hinges.

PROOF. By (4) there is a triangle $(\bar{\gamma}_1, \bar{\tau}, \bar{\sigma})$ in $M^{H-\epsilon}$ with

$$L[\bar{\gamma}_1] = L[\gamma_1], \qquad L[\bar{\tau}] = L[\tau], \qquad L[\bar{\sigma}] = L[\sigma],$$
$$\sphericalangle(-\bar{\gamma}_1'(l_1), \bar{\tau}'(0)) = \bar{\alpha} \le \alpha.$$

Also,

$$\sphericalangle(-\bar{\tau}'(l), -\bar{\sigma}'(k)) = \bar{\alpha}_1 \le \frac{1}{2}\pi.$$

Let $\bar{\theta} : [l_1, l_2] \to M^{H-\epsilon}$ be defined by

$$\bar{\theta}(l) = \bar{\tau}(l), \qquad \bar{\theta}'(l) = \bar{\tau}'(l),$$

and set $\bar{\gamma}_2 = \bar{\tau} \cup \bar{\theta}$. Then

$$\sphericalangle(-\bar{\sigma}'(k), \bar{\theta}'(l)) = \pi - \bar{\alpha}_1 \geq \frac{1}{2}\pi.$$

Since $(\sigma, \theta, \frac{1}{2}\pi)$ is a thin right hinge, it follows by (1),(5) that

$$\rho(\bar{\sigma}(0), \bar{\theta}(l_2)) \geq \rho(\sigma(0), \theta(l_2)).$$

In other words,

$$\rho(\bar{\gamma}_1(0), \bar{\gamma}_2(l_2)) \geq \rho(\gamma_1(0), \gamma_2(l_2)).$$

Since

$$\sphericalangle(-\bar{\gamma}_1'(l_1), \bar{\gamma}_2'(0)) = \bar{\alpha} \leq \alpha.$$

the claim clearly follows by use of (1). □

Define a triangle $(\gamma_1, \gamma_2, \gamma_3)$ to be *thin* if $(\gamma_1, \gamma_2, \alpha_3)$ and $(\gamma_3, \gamma_2, \alpha_1)$ are thin hinges. Then it follows from (5)-(7) and (2) that the theorem holds for thin triangles. We will now prove it in general. Given an arbitrary hinge $(\gamma_1, \gamma_2, \alpha)$ as in (B), fix N and let

$$\tau_{k,l} = \gamma_2 | [\frac{kl_2}{N}, \frac{(k+l)l_2}{N}],$$

where k, l are integers with $0 \leq k, l \leq N$. Let σ_k be minimal from $\gamma_1(0)$ to $\gamma_2(\frac{kl_2}{N})$, and let $T_{k,l} = (\sigma_k, \tau_{k,l}, \sigma_{k+l})$. (See Fig. 2.4). We may suppose

$$L[\gamma_1] + L[\sigma_N] \geq L[\gamma_2],$$

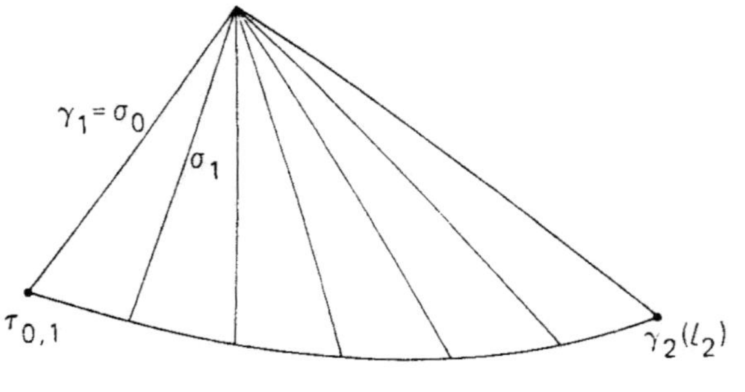

Fig. 2.4

as otherwise (B) follows trivially. We claim that $T_{k,l}$ actually forms a triangle, i.e., all triangle inequalities are satisfied. In fact, we have

$$L[\gamma_1] + L[\sigma_N] \geq L[\gamma_2],$$
$$L[\tau_{0,k}] + L[\sigma_k] \geq L[\gamma_1],$$
$$L[\tau_{k+l,N-k-l}] + L[\sigma_{k+l}] \geq L[\sigma_N].$$

Therefore,

$$\begin{aligned} L[\tau_{0,k}] + L[\sigma_k] + L[\tau_{k+l,N-k-l}] + L[\sigma_{k+l}] &\geq L[\gamma_2] \\ &= L[\tau_{0,k}] + L[\tau_{k,l}] + L[\tau_{k+l,N-k-l}] \end{aligned}$$

or

$$L[\sigma_k] + L[\sigma_{k+l}] \geq L[\tau_{k,l}].$$

Moreover, if N is sufficiently large, an easy compactness argument shows that the triangles $(\sigma_k, \tau_{k,l}, \sigma_{k+l})$ are all *thin*.

For the sake of clarity, there are two points we wish to emphasize. First, nowhere in the definition of thin triangle is it assumed that σ_s, σ_{s+1} be close to one another. Second, we are using the fact that we are working in $M^{H-\epsilon}$. For if we were working in M^H and for some k, $L[\sigma_k] = \frac{\pi}{\sqrt{H}}$, then we could not necessarily guarantee that say $(\sigma_k, \tau_{k,l}, \sigma_{k+l})$ was a thin triangle no matter how short we assumed the length of $\tau_{k,l}$ to be; see the definition of thin right triangle.

(8) (A) is true for $T_{l,k}$ for fixed k and all l implies (B) is true for $T_{l,k+1}$ for fixed k and all l.

PROOF. Assuming (A), there is a triangle $\bar{T}_{l,k}$ in $M^{H-\epsilon}$ with sides $(\bar{\sigma}_k, \bar{\tau}_{k,l}, \bar{\sigma}_{k+l})$ congruent to $T_{l,k}$ and with $\bar{\alpha}_k \leq \alpha_k$ and $\bar{\beta}_{k+l} \leq \beta_{k+l}$, where $\alpha_k = \sphericalangle(\sigma'_k, \tau'_{l,k})$ and $\beta_{k+l} = \sphericalangle(\sigma'_{k+l}, -\tau'_{l,k})$. Note that $\beta_{k+l} + \alpha_{k+l} = \pi$. Extend $\bar{\tau}_{l,k}$ by adding on a segment $\bar{\tau}_{k+l,k+l+1}$ of length $L[\tau_{k+l,k+l+1}]$. Then an argument completely analogous to that in (7) completes the proof. □

(9) It now follows by induction and (3) that (A) and (B) are true for $(\gamma_1, \gamma_2, \gamma_3)$ and $M^{H-\epsilon}$.

(10) Now we have proved (A) and (B) when comparing to triangles in $M^{H-\epsilon}$. In that situation, we know that for all $\epsilon > 0$,

$$\rho(\bar{\gamma}_1^\epsilon(0), \bar{\gamma}_2^\epsilon(l_2)) \geq \rho(\gamma_1(0), \gamma_2(l_2)),$$

where $\bar{\gamma}_1^\epsilon, \bar{\gamma}_2^\epsilon \subset M^{H-\epsilon}$ and make an angle of α_1. The function on the left is clearly continuous in ϵ, so letting $\epsilon \to 0$ yields (B). Although we only showed that (B) implies (A) in case $\epsilon > 0$, a limiting argument similar to the above now also yields (A). Alternatively an appropriate modification of the arguments of (3) may be used. In case $l_i = \frac{\pi}{\sqrt{H}}$ for some i, the triangle obtained in M^H is not unique. □

Recently, Gromoll has pointed out that by means of a somewhat generalized version of Rauch II, steps (4)-(7) of the above argument can be handled simultaneously, thereby shortening the argument considerably. In particular, Rauch I is not used at all and can be deduced as a corollary.

We remark that the assumption that γ_1 is minimal is really necessary. For consider γ_1, γ_2 in $M^{H+\epsilon}$ (ϵ small, $H > 0$) making angle $\alpha > 0$ and of length $\frac{\pi}{\sqrt{H}}$. Their points are at nonzero distance, while the corresponding segments $\bar{\gamma}_1, \bar{\gamma}_2$ in M^H end at the same point.

It is interesting to see what else can be said if the estimates of the theorem actually are equalities. The case $H > 0$ in which M has diameter $\frac{\pi}{\sqrt{H}}$ is treated in Chapter 6. Therefore we will assume that this does not happen.

COROLLARY 2.3. *In the situation of (B) above, suppose* $0 < \alpha < \pi$, $d(M) < \frac{\pi}{\sqrt{H}}$ *and* $\rho(\gamma_1(0), \gamma_2(l_2)) = \rho(\bar{\gamma}_1(0), \bar{\gamma}_2(l_2))$. *Let* $\bar{\gamma}_3$ *denote the unique minimal geodesic in* M^H *from* $\bar{\gamma}_1(0)$ *to* $\bar{\gamma}_2(l_2)$. *Let* Δ *denote the lift to the tangent space at* $\bar{\gamma}_1(0)$ *of the region bounded by the triangle* $(\bar{\gamma}_1, \bar{\gamma}_2, \bar{\gamma}_3)$ *and let* $I : M^H_{\bar{\gamma}_1(0)} \to M_{\gamma_1(0)}$ *be an isometry such that* $I(\bar{\gamma}_i'(0)) = \bar{\gamma}_i'(0)$, $i = 1, 2$. *Then the interior of* $\exp_{\gamma_1(0)} I(\Delta)$ *is a smooth totally geodesic embedded surface of constant curvature H. The image of a minimal geodesic from* $\gamma_1(l_1)$ *to* $\gamma_2(t)$, $0 \le t \le l_2$, *is minimal in M.*

PROOF. We will merely give an indication of the argument. The details will be left to the reader. First of all, since $d(M) < \frac{\pi}{\sqrt{H}}$, Rauch II may be applied to compare thin triangles with corresponding triangles in M^H, rather than $M^{H-\epsilon}$. An examination of the proof of step (8) above shows that all inequalities of that step must be equalities. Then by Corollary (1.36) of Rauch II, the minimal geodesic σ_1 may be chosen to be of the form $\exp_{\gamma_1(l)} f(t)E(t)$ with $E(t)$ parallel, $E(0) = \gamma_2'(0)$ and $f(t)$ as in M^H. Repeated application of that corollary to the thin triangles $T_{l,1}$ and induction shows that σ_{l+1} may also be chosen in this manner. This constructs a totally geodesic region of curvature H spanning γ_1, γ_2. It is easy to see that this region has the description $\exp_{\bar{\gamma}_1(0)} I(\Delta)$. Since the region may be described as a union of minimal geodesics from $\gamma_1(0)$ to $\gamma_2(t)$, $0 \le t \le l_2$ (as follows easily from its construction), the interior must be embedded. For it is clear from the construction that the interior is immersed, and if two such geodesics had an interior point in common, then both could not be minimal (see Chapter 5). Therefore the interior is embedded. □

For purposes of illustration we will give a typical application of Toponogov's Theorem. We will encounter various others in later chapters. The question we want to look at involves the behavior of a geodesic parameterized by arclength $\gamma : [0, \infty) \to M$, where M is a noncompact complete Riemannian manifold. Specifically, we want to know under what conditions does $\lim_{t\to\infty} \rho(\gamma(0), \gamma(t)) = \infty$. Let us look at some examples.

EXAMPLE 2.4 (the cylinder; see Fig. 2.5). There are three classes of geodesics on a cylinder;

(a) the "lines" parallel to the axis;
(b) the closed geodesics perpendicular to the lines;
(c) all the others.

Notice that type (b) are the only geodesics on this noncompact manifold which stay in a compact set for all $0 \le t < \infty$. Also, the geodesics of type

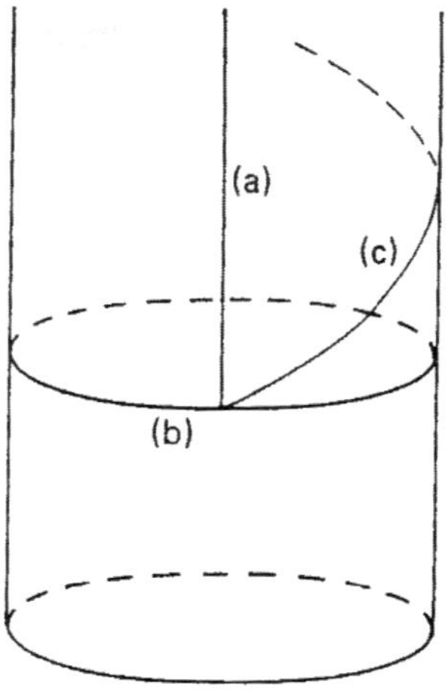

Fig. 2.5.

(a) are the only ones that realize the distance between their endpoints for all time, i.e. $\rho(\gamma(0), \gamma(t)) = t$ for all t.

In an arbitrary noncompact complete Riemannian manifold we define a *ray* to be a geodesic $\gamma : [0, \infty) \to M$ such that $\rho(\gamma(0), \gamma(t)) = t$ for all t. Given $p \in M$, we may always find a ray such that $\gamma(0) = p$. Since M is noncompact, there exists a sequence p_i such that $\rho(p, p_i) \to \infty$. Let γ_i be a sequence of minimal geodesics from p to p_i (which exists by completeness), and let v be an accumulation point of the sequence $\gamma_i'(0)$. Let γ be the geodesic such that $\gamma(0) = p$ and $\gamma'(0) = v$. Then it is easy to see that γ is a ray. In particular, through each p there is a geodesic which eventually leaves every bounded set. Of course, if M is not complete, this assertion is false.

In general, it is not the case that a geodesic must be closed in order to stay in a compact set, as we will now see.

EXAMPLE 2.5 (Weinstein). Let us consider a band B of width $\frac{\pi}{2}$ around the equator of the unit sphere. (See Fig. 2.6). Let $\tilde{B}$ denote the universal covering space of B with its natural induced Riemannian metric. $\tilde{B}$ may be described as $\mathbb{R} \times [-\frac{1}{4}\pi, \frac{1}{4}\pi]$ with metric $\cos^2 y\,dx^2 + dy^2$, as the reader can easily check. It has a group of isometries I which is naturally isomorphic to $\mathbb{R}$, where $I \ni I_t : (x, y) \to (x + t, y)$ and $B = \tilde{B}/\{I_{2\pi}\}$. Consider $\hat{B} = \tilde{B}/\{I_{2\pi\alpha}\}$, where α is an irrational number between 0 and 1. The inverse image $\tilde{C}$ of a great circle C which makes a sufficiently small angle with the equator stays in $\tilde{B}$ for all time. The image $\hat{C}$ of $\tilde{C}$ in $\hat{B}$ is easily seen not to be periodic (as in Fig. 2.7). Now complete $\hat{B}$ by pulling out its boundary to infinity.

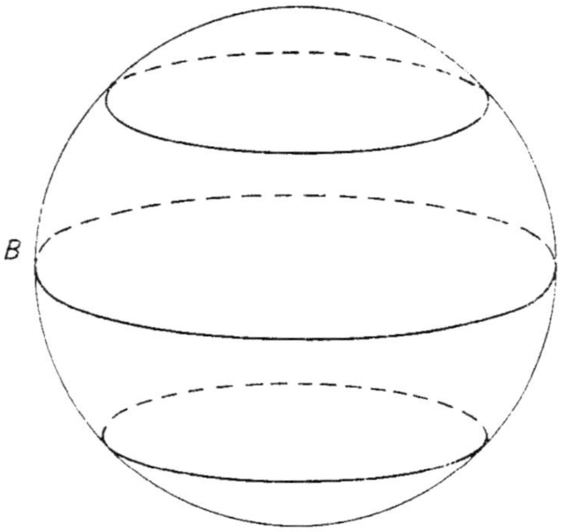

Fig. 2.6.

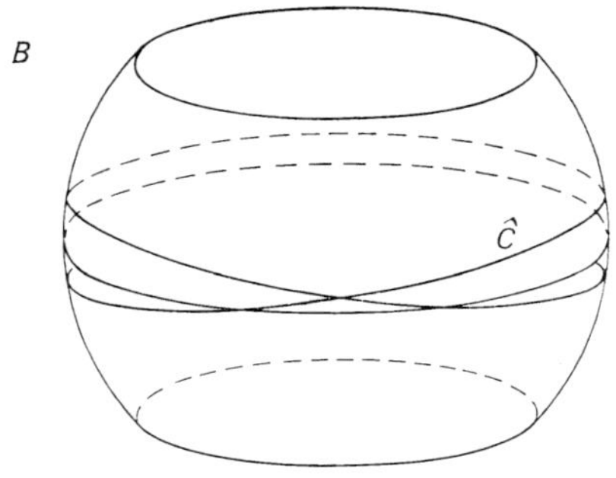

Fig. 2.7.

More formally, consider $\mathbb{R} \times \mathbb{R}$ with metric $\rho(y)\cos^2(y)\mathrm{d}x^2 + \mathrm{d}y^2$, where $\rho(y)$ is a smooth function equal to 1 on $[-\frac{1}{4}\pi, \frac{1}{4}\pi]$ and equal to $\cos^{-2} y$ outside of $(-\frac{3}{8}\pi, \frac{3}{8}\pi)$. $\tilde{C}$ remains in the band $-\frac{1}{4}\pi \le y \le \frac{1}{4}\pi$, and we may again divide out by $\{I_{2\pi\alpha}\}$. (See Fig. 2.8).

With these examples in mind, especially Example 2.4, we state a theorem which gives some information in the case where M has nonnegative sectional curvature.

THEOREM 2.6. *Let M be complete and have nonnegative sectional curvature, and let $\gamma, \sigma : [0, \infty) \to M$ be geodesics such that $\gamma(0) = \sigma(0)$. If γ is a ray and $\sphericalangle(\gamma'(0), \sigma'(0)) < \frac{1}{2}\pi$, then $\lim_{t\to\infty} \rho(\sigma(0), \sigma(t)) = \infty$ (or more briefly, σ goes to ∞).*

The idea of the proof is basically this: γ goes to infinity as quickly as possible, and σ makes an acute angle with γ. Since the nonnegative

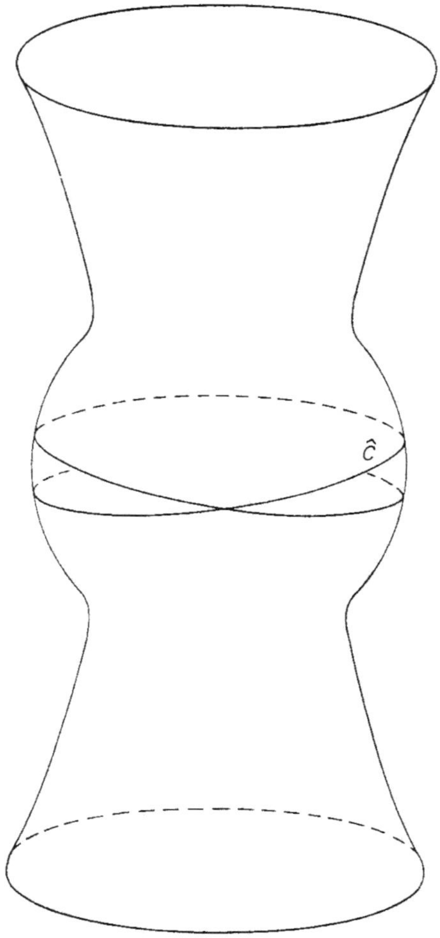

Fig. 2.8.

curvature of M tends to pull σ towards γ, it is reasonable that the behavior of σ should in some rough sense approximate the behavior of γ.

Example 2.4 shows that $< \frac{1}{2}\pi$ cannot be changed to $\leq \frac{1}{2}\pi$. However, in Theorem 8.24 we shall see that if the curvature of M is strictly positive, *all* geodesics $\sigma : [0, \infty) \to M$ go to ∞. Therefore, we may certainly replace $< \frac{1}{2}\pi$ by $\leq \frac{1}{2}\pi$ in this case. We are led to conjecture that if M has nonnegative sectional curvature, $\sphericalangle(\gamma'(0), \sigma'(0)) = \frac{1}{2}\pi$, and σ does not go to ∞, then $\gamma'(0), \sigma'(0)$ should span a section of curvature zero. Actually, something more is true, and this rigidity phenomenon is also discussed in Theorem 8.24. Finally, the theorem is completely false without the assumption that M has nonnegative curvature, as an inspection of Example 2.5 shows.

PROOF. By assumption $\sphericalangle(\gamma'(0), \sigma'(0)) = \alpha < \frac{1}{2}\pi$. Since by the triangle inequality we have

$$\rho(\sigma(t), \sigma(0)) \geq \rho(\gamma(s), \sigma(t)) - \rho(\gamma(s), \gamma(0)),$$

it will suffice to show that the limit of the right-hand side as $s \to \infty$ and for fixed t is $\geq t \cos \alpha$.

Consider the triangle formed by $\sigma|[0, t], \gamma|[0, s]$ and $\tau_{s,t}$, where $\tau_{s,t}$ is a minimal segment from $\sigma(t)$ to $\gamma(s)$. $\sigma|[0, t]$ is of course not necessarily minimal, but Toponogov's Theorem (A) still applies.

Thus we can construct a congruent triangle in $\mathbb{R}^2$ with no bigger angles. Let α denote the angle at $\gamma(0)$ and $\alpha^* \leq \alpha$ the corresponding angle in $\mathbb{R}^2$. By the law of cosines in $\mathbb{R}^2$ we have

$$t^2 + s^2 = L[\tau_{s,t}]^2 - 2ts \cos \alpha^*.$$

Therefore,

$$L[\tau_{s,t}] - s = \frac{t^2 + 2ts \cos \alpha^*}{s + L[\tau_{s,t}]}.$$

But as $s \to \infty$, by the triangle inequality, $\frac{L[\tau_{s,t}]}{s} \to 1$. Hence $(L[\tau_{s,t}] - s)$ approaches $t \cos \alpha^* \geq t \cos \alpha$. $\square$

CHAPTER 3

Homogeneous Spaces

In this chapter we shall study invariant metrics on Homogeneous spaces – spaces on which a Lie group acts transitively. Homogeneous spaces are, in a sense, the nicest examples of Riemannian manifolds and are good spaces on which to test conjectures.

We shall need some elementary facts about Lie groups, which we shall summarize without proof. The reader who is not familiar with this material should consult Chevalley [1946], Helgason [1962], Sternberg [1964]. We shall also use Frobenius' Theorem and various properties of the Lie derivative. (See Sternberg [1964]).

DEFINITION 3.1. A *Lie group* G is a smooth manifold (which we do not assume connected), which has the structure of a group in such a way that the map $\phi : G \times G \to G$ defined by $\phi(x, y) = x.y^{-1}$ is smooth.

It can be shown that a C^∞ Lie group has a compatible real analytic structure (see Chevalley [1946]). Canonically associated to a Lie group is its Lie algebra.

DEFINITION 3.2. A *Lie algebra* is a vector space V together with a map $[\ ,\ ] : V \times V \to V$ such that

(1) $[a_1V_1 + a_2V_2, W] = a_1[V_1, W] + a_2[V_2, W]$

(2) $[V, W] = -[W, V]$

(3) $[V_1, [V_2, V_3]] + [V_2, [V_3, V_1]] + [V_3, [V_1, V_2]] = 0.$

The last relation is called the Jacobi identity.

EXAMPLE 3.3. If M is a smooth manifold, then $\chi(M)$ is a Lie algebra (of infinite dimension) with respect to the bracket operation $[X, Y](f) = (XY - YX)f$. To check the Jacobi identity is straightforward:

$$\begin{aligned}
&[X, [Y, Z]] + [Y, [Z, X]] + [Z, [X, Y]] \\
&\quad = X(YZ - ZY) - (YZ - ZY)X + Y(ZX - XZ) \\
&\qquad - (ZX - XZ)Y + Z(XY - YX) - (XY - YX)Z \\
&\quad = XYZ - XZY - YZX + ZYX + YZX - YXZ - ZXY + XZY \\
&\qquad + ZXY - ZYX - XYZ + YXZ = 0.
\end{aligned}$$

We shall now describe the Lie algebra associated with Lie group G. If G is a Lie group, we have for each $g \in G$ the diffeomorphisms $L_g : g_1 \to gg_1$ and $R_g : g_1 \to g_1g$. We say that $V \in \chi(G)$ is *left invariant* (respectively *right invariant*) if $dL_g(V(g_1)) = V(gg_1)$ (respectively $dR_g(V(g_1)) = V(g_1g)$. If V is left invariant, then it is uniquely determined by $V(e)$, where e is the identity element of G.

Conversely $V \in G_e = \mathfrak{g}$ gives rise to a left invariant vector field (l.i.v.f.) $V(g) = dL_g(V(e))$. Since multiplication on G is smooth, so is a l.i.v.f. Therefore the l.i.v.f.'s form an n-dimensional subspace of $\chi(G)$, and we claim that the bracket of two l.i.v.f.'s is again left invariant. In fact it follows from the definition of Lie bracket that for any diffeomorphism $\phi : M \to M$ and $X, Y \in \chi(M)$,

$$d\phi[X,Y] = [d\phi(X), d\phi(Y)].$$

Then

$$dL_g[X,Y] = [dL_g(X), dL_g(Y)] = [X,Y]$$

if X, Y are l.i.v.f.'s. It follows that the l.i.v.f.'s form a Lie algebra $\mathfrak{g}$, the *Lie algebra of G*. Of course the choice of l.i.v.f. rather than r.i.v.f. is only a convention. The r.i.v.f.'s also form a Lie algebra isomorphic to the l.i.v.f.'s. It is often convenient to identify $\mathfrak{g}$ with G_e as above, and we will use both interpretations simultaneously. We note that as a consequence of this discussion, it follows that the tangent bundle of G is trivial. If $\bar{X}$ denotes the r.i.v.f. such that $\bar{X}(e) = X(e)$ for some l.i.v.f. X, then we shall see in Proposition 3.7 that

$$[\bar{X}, \bar{Y}]|_e = [-\bar{X}, -\bar{Y}]|_e = \overline{[X,Y]}|_e.$$

Hence the map $X \to -\bar{X}$ induces the isomorphism between the two Lie algebras.

PROPOSITION 3.4. *Let $\phi : G_1 \to G_2$ be a continuous homomorphism of Lie groups. Then ϕ is a real analytic map and hence induces* $d\phi$, *which is a homomorphism of Lie algebras.*

It can be shown that if G_1 is simply connected and $f : G_{e_1} \to G_{e_2}$ is a homomorphism of Lie algebras, then there exists a unique analytic homomorphism $\phi : G_1 \to G_2$ such that $d\phi = f$. Also, any finite dimensional Lie algebra is the Lie algebra of a simply connected Lie group. In this way the classification of simply connected Lie groups can be reduced to the algebraic problem of classification of Lie algebras. In case $\mathfrak{g}$ is semisimple (defined before Proposition 3.39), this classification can be carried out explicitly. Finally, if G is a Lie group, any covering space of G is a Lie group in a natural way with Lie algebra isomorphic to $\mathfrak{g}$.

A subspace $\mathfrak{h}$ of $\mathfrak{g}$ which is closed under $[\ ,\ ]$ is called a *subalgebra* of $\mathfrak{g}$. A subalgebra $\mathfrak{J}$ such that $[x, \mathfrak{J}] \subset \mathfrak{J}$ for all $x \in \mathfrak{J}$ is called an *ideal.*

PROPOSITION 3.5. *If $\mathfrak{h}$ is a subalgebra of $\mathfrak{g}$ (considered as a Lie algebra of vector fields), then $\mathfrak{h}$ defines a involutive distribution and the maximal connected integral manifold H through e is a subgroup (which will not, in* *general, be a closed subset of G).*[1] *Conversely, if $H \subset G$ is a Lie subgroup (a subgroup which is also a 1-1 immersed submanifold), then the tangent space $\mathfrak{h}$ to H at e is a subalgebra of $\mathfrak{g}$. H is a normal subgroup if and only if $\mathfrak{h}$ is an ideal.*

As a special case of Proposition 3.5, we may take $\mathfrak{h}$ to be any 1-dimensional subspace of $\mathfrak{g}$. Then $[\mathfrak{h}, \mathfrak{h}] = 0 \subset \mathfrak{h}$. The subgroup corresponding to such an $\mathfrak{h}$ is called a *one-parameter subgroup*. For any $v \in G_e = \mathfrak{g}$, we have a natural homomorphism of Lie algebras $\mathrm{d}\phi : \mathbb{R} \to \mathfrak{g}$ with $\mathrm{d}\phi(1) = v$, and hence a Lie group homomorphism $\phi : \mathbb{R} \to G$ mapping $\mathbb{R}$ onto the integral curve through the origin of the l.i.v.f. corresponding to V. We denote $\phi(1)$ by e^v. Then $e^{t_1 v} e^{t_2 v} = e^{(t_1+t_2)v}$. Moreover, there exists a neighborhood U of $0 \in G_e$ such that $e : U \to G$ is a diffeomorphism onto a neighborhood of the identity on G.

PROPOSITION 3.6. *If $\phi : G_1 \to G_2$ is a homomorphism, then $\phi(e^v) = e^{\mathrm{d}\phi(v)}$.*

The following proposition will also be useful.

PROPOSITION 3.7.

$$\frac{\partial}{\partial t}\frac{\partial}{\partial s}\left(e^{tx}e^{sy}e^{-tx}\right)\Big|_{t=0,s=0} = [X,Y]|_e = -[\bar{X},\bar{Y}]|_e.$$

A word about the precise meaning of the expression in Proposition 3.7. For each fixed t, $e^{tx}e^{sy}e^{-tx}$ is a curve through the origin.

$$\frac{\partial}{\partial s}\left(e^{tx}e^{sy}e^{-tx}\right)\Big|_{s=0}$$

is then a tangent vector in G_e. As we let t vary,

$$\frac{\partial}{\partial s}\left(e^{tx}e^{sy}e^{-tx}\right)\Big|_{s=0}$$

describes a curve in G_e. It makes sense to differentiate this curve at $t = 0$ and the result is a tangent vector in G_e.

PROOF. Let ϕ_t be the one-parameter group of diffeomorphisms generated by X. Then by the alternative definition of Lie bracket (see Sternberg [1964]),

$$[X,Y] = \frac{\mathrm{d}}{\mathrm{d}t}\mathrm{d}\phi_{-t}(y(\phi_t(e)))\Big|_{t=0}$$

Now the integral curve of Y through $\phi(t)$ is equal to $e^{tx}e^{sy}$ by the left invariance of Y. Then

$$[X,Y] = \frac{\partial}{\partial t}\frac{\partial}{\partial s}\phi_{-t}(e^{tx}e^{sy})\Big|_{t=0,s=0}$$

[1] H is in general only a 1-1 immersed submanifold, so its manifold topology is not always the relative topology.

But for arbitrary $g \in G$, $\phi_{-t}(g)$ is by definition the endpoint of the integral curve of X through g parameterized on the interval $[0,-t]$. By left invariance of the integral curves of X we then have

$$\phi_{-t}g = ge^{-tx} = R_{e^{-tx}}(g).$$

Hence

$$[X,Y] = \frac{\partial}{\partial t}\frac{\partial}{\partial s}\left(e^{tx}e^{sy}e^{-tx}\right)\Big|_{t=0,s=0}.$$

The other equation follows similarly. □

EXAMPLE 3.8. Let $M^n = \mathbb{R}^{n^2}$ denote the space of $n \times n$ matrices

$$GL(n) = \{m \in M \big| \det m \neq 0\}.$$

Since $\det(m_1 \times m_2) = \det(m_1)\det(m_2)$ and $\det(I) = 1$, it follows that $GL(n)$ is a group. $GL(n)$ is in a natural way an open subset of $\mathbb{R}^{n^2}$ and hence a manifold. Define

$$e^{tm} = I + tm + t^2\frac{m^2}{2!} + \ldots$$

This series can easily be shown to converge for all t to a continuous function satisfying

$$e^{t_1 v}e^{t_2 v} = e^{(t_1+t_2)v}.$$

Moreover, for any m,

$$e^{tr(m)} = \det(e^m)$$

(as follows easily using the Jordan canonical form). It follows that the $\{e^{tm}\}$ are the one-parameter subgroups of $GL(n)$, and the Lie algebra gl(n) of $GL(n)$ is naturally M^n. Now by Proposition 3.7,

$$\begin{aligned}
[m_1,m_2] &= \frac{\partial}{\partial t}\frac{\partial}{\partial s}\left(e^{tm_1}e^{sm_2}e^{-tm_1}\right)\Big|_{t=0,s=0} \\
&= \frac{\partial}{\partial t}\frac{\partial}{\partial s}\left\{(I + tm_1 + \ldots)(I + sm_2 + \ldots)(I - tm_1 + \ldots)\right\}\Big|_{t=0,s=0} \\
&= \frac{\partial}{\partial t}\frac{\partial}{\partial s}\left(I + ts(m_1m_2 - m_2m_1) + \ldots\right)\Big|_{t=0,s=0} \\
&= m_1m_2 - m_2m_1.
\end{aligned}$$

We refer the reader to Chevalley [1946] for the description of the standard matrix subgroups of $GL(n)$.

EXAMPLE 3.9. Myer and Steenrod [1939] have shown that the isometry group of any Riemannian manifold is a Lie group. We will be using this fact implicitly below.

Given a Lie group G of dimension n, there is a natural homomorphism, the *adjoint representation*, from G to $GL(\mathfrak{g}) \simeq GL(n)$, defined as follows.

DEFINITION 3.10. $\mathrm{Ad}_g(x) = \mathrm{d}R_g \circ \mathrm{d}L_{g^{-1}}(x)$.

Clearly $\mathrm{Ad}_{g_1 g_2} = \mathrm{Ad}_{g_1}\mathrm{Ad}_{g_2}$. Since for each g, the map $h \to ghg^{-1}$ is an automorphism of G, it follows that Ad_g is an automorphism of $\mathfrak{g}$,

$$\mathrm{Ad}_g([x,y]) = [\mathrm{Ad}_g(x), \mathrm{Ad}_g(y)].$$

We set $\mathrm{ad} = \mathrm{d}(\mathrm{Ad})$. In other words $\mathrm{ad} : \mathfrak{g} \to \mathrm{gl}(\mathfrak{g})$ is the differential of the adjoint representation. We see from Proposition 3.7 that

$$\mathrm{ad}_x(y) = \frac{\partial}{\partial t}\frac{\partial}{\partial s}\left(e^{tx}e^{sy}e^{-tx}\right)\Big|_{t=0,s=0} = [x,y]. \tag{3.11}$$

From the Jacobi identity it follows that ad_x is a derivation,

$$\mathrm{ad}_x[y,z] = [\mathrm{ad}_x(y), z] + [y, \mathrm{ad}_x(z)].$$

From the remark after Proposition 3.5 it follows that $\mathrm{Ad}_{e^x} = e^{\mathrm{ad}_x}$.

We will now begin the study of homogeneous spaces.

DEFINITION 3.12. If G is a connected Lie group and H a closed subgroup, G/H is the space of cosets $\{gH\}$, $\pi : G \to G/H$ is defined by $g \to [gH]$. G/H is called a *homogeneous space.*

Notice that for $h \in G$ we have $h(H) = [H]$ if and only if $h \in H$.

PROPOSITION 3.13. *A subgroup which is a closed subset is an analytic sub-manifold and hence a Lie subgroup.*

PROPOSITION 3.14. *G/H has a unique real analytic structure for which $\pi : G \to G/H$ is an analytic fibration.*

There is a natural smooth left action of G on G/H defined by $g_1[gH] = [g_1 gH]$. The diffeomorphism, L_g, of G/H induced by g will sometimes be denoted g. Since $g_1 g^{-1}[gH] = [g_1 H]$, the action of G is transitive; hence the terminology, homogeneous space. We want to study metrics on G/H for which G acts by isometries. Such metrics are called *invariant.* In G itself we may also consider the right action of G. Metrics invariant under both left and right actons are called *bi-invariant.* We should point out that invariant metrics do not exist for all G/H. Moreover, when they do exist, G may not be the full group $\hat{G}$ of isometries. $\hat{H}$, the largest group of isometries fixing some point $[gH] \in G/H$, is called the *isotropy group* of that point. $\hat{H}$ is identified with closed subgroup of the orthogonal group of $G/H_{[H]}$, and hence is compact. This identification comes from the fact that on a connected Riemannian manifold an isometry is determined by its differential at a single point. It is easy to verify this by using the fact that isometries commute with the exponential map. That $\hat{H}$ is closed follows from the Cartan-Ambrose-Hicks Theorem.

G is said to act *effectively* on G/H if $L_g = 1$ (the identity map) implies $g = e$. Let H_0 be the largest subgroup of H which is normal in G. Set

$$G^* = G/H_0, \qquad H^* = H/H_0.$$

Then it is straightforward to check that G^*/H^* is diffeomorphic to G/H and that G^* acts effectively on G^*/H^*. This reduction may require some work in dealing with a specific example. If G acts effectively, then it may be identified with a Lie subgroup of $\hat{G}$. (Again this identification is 1-1 but not always an embedding.) In this case it is not hard to see that

$$\dim \hat{G} - \dim G = \dim \hat{H} - \dim H.$$

The tangent space to the point $[H]$ of G/H can be naturally identified with $\mathfrak{g}/\mathfrak{h}$. Further, since the action of G on G and G/H commutes with π, we have for $h \in H$ and $v \in \mathfrak{g}$

$$he^{tv}H = he^{tv}h^{-1}H.$$

Ad_H and $\mathrm{ad}_{\mathfrak{h}}$ leave $\mathfrak{h}$ invariant and hence act naturally on $\mathfrak{g}/\mathfrak{h}$. Differentiating with respect to t yields

$$\mathrm{d}L_h(v) = \pi\Big(\mathrm{Ad}_h(v)\Big), \tag{3.15}$$

where $\pi : \mathfrak{g} \to \mathfrak{g}/\mathfrak{h}$ is the natural projection.

PROPOSITION 3.16. *(1) The set of G-invariant metrics on G/H is naturally isomorphic to the set of scalar products* $\langle\ ,\ \rangle$ *on* $\mathfrak{g}/\mathfrak{h}$ *which are invariant under the action of* Ad_H *on* $\mathfrak{g}/\mathfrak{h}$.

(2) If H is connected, a scalar product $\langle\ ,\ \rangle$ *is invariant under* Ad_H *if and only if for each* $h \in \mathfrak{h}$, ad_h *is skew symmetric with respect to* $\langle\ ,\ \rangle$.

(3) If G acts effectively on G/H, then G/H admits a G-invariant metric if and only if the closure $\mathrm{cl}(\mathrm{Ad}_H)$ *of the group* $\mathrm{Ad}_H \subset GL(\mathfrak{g})$ *is compact.*

(4) If G acts effectively on G/H, and if $\mathfrak{g}$ *admits a decomposition* $\mathfrak{g} = \mathfrak{p} \bigoplus \mathfrak{h}$ *with* $\mathrm{Ad}_H(\mathfrak{p}) \subset \mathfrak{p}$ *then G-invariant metrics on G/H are in 1-1 correspondence with* Ad_H*-invariant scalar products on* $\mathfrak{p}$. *These exist if and only if the closure of the group* $\mathrm{Ad}_H|\mathfrak{p}$ *is compact. Conversely, if G/H admits a G-invariant metric, then G admits a left invariant metric which is right invariant under H, and the restriction of this metric to H is bi-invariant.*

Setting $\mathfrak{p} = \mathfrak{h}^{\perp}$ *gives a decomposition as above.*

(5) If H is connected, the condition $\mathrm{Ad}_H(\mathfrak{p}) \subset \mathfrak{p}$ *is equivalent to* $[\mathfrak{h}, \mathfrak{p}] \subset \mathfrak{p}$.

(6) If G is compact, then G admits a bi-invariant metric.

PROOF. (1) Given a left invariant metric on G/H, by restricting to the tangent space at $[H]$ we get an inner product on $\mathfrak{g}/\mathfrak{h}$. By (3.15), $\langle\ ,\ \rangle$ is invariant under Ad_H. Conversely, given such a $\langle\ ,\ \rangle$ we get an inner product on $G/H_{[H]}$. Given $[gH]$ we may define an inner product $\langle\ ,\ \rangle_{[H]}$ on $G/H_{[gH]}$ by setting

$$\langle x, y\rangle_{[gH]} = \langle \mathrm{d}L_{g^{-1}}(x), \mathrm{d}L_{g^{-1}}(y)\rangle_{[H]}.$$

Since

$$\begin{aligned}\langle \mathrm{d}L_{hg^{-1}}(x), \mathrm{d}L_{hg^{-1}}(y)\rangle_{[H]} &= \langle \mathrm{d}L_h \mathrm{d}L_{g^{-1}}(x), \mathrm{d}L_h \mathrm{d}L_{g^{-1}}(y)\rangle_{[H]} \\ &= \langle \mathrm{d}L_{g^{-1}}(x), \mathrm{d}L_{g^{-1}}(y)\rangle_{[H]}\end{aligned}$$

if $\langle\ ,\ \rangle$ is invariant under Ad_H, $\langle\ ,\ \rangle_{[gH]}$ is independent of which member of $[gH]$ we chose to define it. In this way we get a Riemannian metric on G/H which is clearly left invariant.

(2) That the condition

$$\langle \mathrm{Ad}_{e^{tv}} x, \mathrm{Ad}_{e^{tv}} y\rangle = \langle x, y\rangle$$

for all $x, y \in \mathfrak{g}/\mathfrak{h}$ and $v \in \mathfrak{h}$ implies

$$\langle \mathrm{ad}_v x, y\rangle + \langle x, \mathrm{ad}_v y\rangle = 0$$

follows from Proposition 3.7. Conversely, if we assume the second condition, then by Proposition 3.6, for all x, y, v,

$$\begin{aligned}\langle \mathrm{Ad}_{e^{tv}} x, \mathrm{Ad}_{e^{tv}} y\rangle &= \langle e^{\mathrm{ad}_{tv}} x, e^{\mathrm{ad}_{tv}} y\rangle = \sum \langle e^{\mathrm{ad}_{tv}} x, (t^n/n!)(\mathrm{ad}_v)^n y\rangle \\ &= \sum \langle (-1)^n (t^n/n!)(\mathrm{ad}_v)^n e^{\mathrm{ad}_{tv}} x, y\rangle = \langle e^{\mathrm{ad}_{-tv}} e^{\mathrm{ad}_{tv}} x, y\rangle.\end{aligned}$$

Now the set of elements for which the claim holds obviously forms a closed subgroup $\bar{H} \subset H$. On the other hand, since every element of some open neighborhood U of the identity is of the form e^{tv}, the claim holds for elements of U. Thus the Lie algebra must be equal to that of H. Since $\bar{H}$ is connected, $\bar{H} = H$.

(3) Let G^*, H^* denote the isometry and isotropy groups of G/H. Since G acts effectively, we have 1-1 homomorphism $G \to G^*$ inducing $\mathfrak{g} \to \mathfrak{g}^*$. The group H^* is compact and therefore so is its image $\mathrm{Ad}_{H^*} : \mathfrak{g}^* \to \mathfrak{g}^*$. Let ω be a right invariant volume form on Ad_{H^*}, coming, for example from a right invariant metric. Then for any inner product $\langle\ ,\ \rangle$ on $\mathfrak{g}^*$, define

$$\langle\langle x, y\rangle\rangle = \int_{\mathrm{Ad}_{H^*}} \langle \mathrm{Ad}_{h^*}(x), \mathrm{Ad}_{h^*}(y)\rangle \omega(h^*).$$

Then Ad_{h^*} acts by isometries with respect to $\langle\langle\ ,\ \rangle\rangle$ because

$$\langle\langle \mathrm{Ad}_{h_1}(x), \mathrm{Ad}_{h_1}(y)\rangle\rangle = \int_{\mathrm{Ad}_{H^*}} \langle \mathrm{Ad}_{h^*}\mathrm{Ad}_{h_1}(x), \mathrm{Ad}_{h^*}\mathrm{Ad}_{h_1}(y)\rangle \omega(h^*).$$

Since Ad is a homomorphism and ω is right invariant, this becomes

$$\begin{aligned}&\int_{\mathrm{Ad}_{H^*}} \langle \mathrm{Ad}_{h^* h_1}(x), \mathrm{Ad}_{h^* h_1}(y)\rangle \mathrm{d}R_{h_1^{-1}}\omega(h^* h_1) \\ &\qquad = \int_{\mathrm{Ad}_{H^*}} \langle \mathrm{Ad}_{h^*}(x), \mathrm{Ad}_{h^*}(y)\rangle \mathrm{d}R_{h_1^{-1}}\omega(h^*).\end{aligned}$$

Since $R_{h_1^{-1}}$ is a diffeomorphism, this becomes

$$\int_{\mathrm{Ad}_{H^*}} \langle \mathrm{Ad}_{h^*}(x), \mathrm{Ad}_{h^*}(y)\rangle \omega(h^*) = \langle\langle x, y\rangle\rangle.$$

Now the restriction of $\langle\langle\ ,\ \rangle\rangle$ to $\mathfrak{g}$ is an inner product with respect to which Ad_H acts by isometries. Hence Ad_H is contained in the (compact) orthogonal group with respect to this inner product, which implies that its closure is compact. Conversely, if the closure of Ad_H is compact, in a manner similar to the above we may construct an inner product $\langle\langle\ ,\ \rangle\rangle$ on $\mathfrak{g}$ such that Ad_H acts by isometries. Let $\mathfrak{p} = \mathfrak{h}^\perp$ with respect to $\langle\langle\ ,\ \rangle\rangle$. Then $\langle\langle\ ,\ \rangle\rangle|\mathfrak{p}$ induces an Ad_H-invariant inner product on $\mathfrak{g}/\mathfrak{h}$ under the identification $\rho : \mathfrak{p} \to \mathfrak{g}/\mathfrak{h}$.

(4) In view of (1), (3) this is straightforward to check.

(5) This follows as in (2).

(6) This proof is similar to that of (3). □

In Proposition 3.34 we will show that a simply connected Lie group H which admits a bi-invariant metric is the product of a compact group and a vector group $\mathbb{R}^k$ which is the center of H.

The following is an example of a homogeneous space which does not admit a left invariant metric.

EXAMPLE 3.17. $SL(n)/SL(n-1)$;
$SL(n) = \{m \in M^n | \det m = 1\}$;
$SL(n-1)$ may be embedded in $SL(n)$ by setting

$$m = \begin{bmatrix} 1 & 0 \\ 0 & m \end{bmatrix}$$

for all $m \in SL(n-1)$.

$SL(n)$ acts effectively on $SL(n)/SL(n-1)$. The reader may verify that $SL(n-1)$ is not the product of a compact group and a vector group (see Proposition 3.34).

We are now going to compute the curvature of a left invariant metric on G/H. We begin with the special case of a left invariant metric on G itself. We use the notation A^* to denote the adjoint of the linear transformation A with respect to a given inner product.

PROPOSITION 3.18. *Let $\langle\ ,\ \rangle$ be a left invariant metric on G and let X, Y, Z be l.i.v.f.'s. Then:*

(1) $\nabla_X Y = \frac{1}{2}\{[X,Y] - (\mathrm{ad}_X)^*(Y) - (\mathrm{ad}_Y)^*(X)\}$;

(2) $\langle R(X,Y)Z, W\rangle = \langle \nabla_X Z, \nabla_Y W\rangle - \langle \nabla_Y Z, \nabla_X W\rangle - \langle \nabla_{[X,Y]} Z, W\rangle$;

(3)

$$\begin{aligned}\langle R(X,Y)Y, X\rangle = {}& \|[(\mathrm{ad}_X)^*(Y) + (\mathrm{ad}_Y)^*(X)]\|^2 \\ & - \langle (\mathrm{ad}_X)^*(X), (\mathrm{ad}_Y)^*(Y)\rangle - \frac{3}{4}\|[X,Y]\|^2 \\ & - \frac{1}{2}\langle [[X,Y],Y], X\rangle - \frac{1}{2}\langle [[Y,X],X], Y\rangle;\end{aligned}$$

(4) One-parameter subgroups are geodesics if and only if $\mathrm{ad}_X^*(X) = 0$ *for all* X.

PROOF. By left invariance we have

$$\begin{aligned} 0 &= X\langle Y, Z\rangle = \langle \nabla_X Y, Z\rangle + \langle Y, \nabla_X Z\rangle, \\ 0 &= Y\langle X, Z\rangle = \langle \nabla_Y X, Z\rangle + \langle X, \nabla_Y Z\rangle, \\ 0 &= Z\langle X, Y\rangle = \langle \nabla_Z X, Y\rangle + \langle X, \nabla_Z Y\rangle. \end{aligned}$$

Computing as in Chapter 1, Section 1 gives

$$\langle \nabla_X Y, Z\rangle = \frac{1}{2}\{\langle [X,Y],Z\rangle - \langle Y,[X,Z]\rangle - \langle X,[Y,Z]\rangle\},$$

from which (1) readily follows.

(2) By left invariance, $X\langle \nabla_Y Z, W\rangle = 0$. Therefore,

$$\begin{aligned}\langle \nabla_X \nabla_Y Z, W\rangle &= -\langle \nabla_Y Z, \nabla_X W\rangle,\\ -\langle \nabla_Y \nabla_X Z, W\rangle &= \langle \nabla_X Z, \nabla_Y W\rangle,\\ -\langle \nabla_{[X,Y]} Z, W\rangle &= -\langle \nabla_{[X,Y]} Z, W\rangle.\end{aligned}$$

(2) follows by adding these equations.

(3) follows from (1) and (2).

(4) is immediate from (1). □

In the case of a bi-invariant metric, (3) above simplifies considerably.

COROLLARY 3.19. *If $\langle\ ,\ \rangle$ is bi-invariant, then:*

(1) $\nabla_X Y = \frac{1}{2}[X,Y]$;

(2) $\langle R(X,Y)Z,W\rangle = \frac{1}{4}(\langle [X,W],[Y,Z]\rangle - \langle [X,Z],[Y,W]\rangle)$;

(3) $\langle R(X,Y)Y,X\rangle = \frac{1}{4}||[X,Y]||^2$. *In particular the sectional curvature is non-negative.*

(4) One-parameter subgroups are geodesics.

PROOF. Since $\langle\ ,\ \rangle$ is bi-invariant, by Proposition 3.16 we have

$$\langle Y,[X,Z]\rangle = -\langle Y,[Z,X]\rangle = \langle [Z,Y],X\rangle.$$

(1) now follows from the proof of (1) of Proposition 3.18.

(2) Substituting (1) into (2) of Proposition 3.18 gives

$$\begin{aligned}\langle R(X,Y)Z,W\rangle = \frac{1}{4}\langle [X,Z],[Y,W]\rangle - \frac{1}{4}\langle [Y,Z],[X,W]\rangle\\ - \frac{1}{2}\langle [[X,Y],Z],W\rangle.\end{aligned}$$

Using the Jacobi identity, the last term may be rewritten as

$$\begin{aligned}-\frac{1}{2}\langle [Y,[Z,X]],W\rangle - \frac{1}{2}\langle [X,[Y,Z]],W\rangle\\ = -\frac{1}{2}\langle [X,Z],[Y,W]\rangle + \frac{1}{2}\langle [Y,Z],[X,W]\rangle,\end{aligned}$$

and (2) follows.

(3) Follows immediately from (2).

(4) One-parameter subgroups are the orbits of l.i.v.f.'s, so (1) implies (4). □

In order to generalize our formulas to the case of an arbitrary homogeneous space, we will prove a formula of O'Neill [1966], on the curvature of Riemannian submersions. A *submersion* is a differentiable map $\pi : M^{n+k} \to N^n$ such that at each point $d\pi$ has rank n. It follows from the implicit-function theorem that $\pi^{-1}(p)$ is a smooth k-dimensional submanifold of M for all $p \in N$. Let V denote the tangent space of $\pi^{-1}(p)$ at $q \in \pi^{-1}(p)$. Assume that M and N have Riemannian metrics and set $H = V^{\perp}$. We call H and V the *horizontal* and *vertical* subspaces, respectively, and we use H and V as superscripts to denote horizontal and vertical components. π is called a *Riemannian submersion* if $\mathrm{d}\pi | H$ is an isometry. If X is a vector field on N, then there is a unique vector field $\bar{X}$ on M such that $\bar{X} \in H$ and $\mathrm{d}\pi(\bar{X}) = X$. Also if $c : [0,1] \to N$ is a piecewise smooth curve, and $q \in \pi^{-1}(c(0))$, then there is a unique curve $\bar{c} : [0,1] \to M$ such that $\bar{c}(0) = q$, and $\pi \circ \bar{c} = c$, $\bar{c}'(t) \in H$. This follows from the theory of ordinary differential equations exactly as in the special case in which M is a principal bundle over N and H defines a connection; see Kobayashi and Nomizu [1963, 1969].

We now give a formula which relates the curvature $\bar{K}(\bar{X}, \bar{Y})$ of a plane section spanned by the orthonormal vectors $\bar{X}, \bar{Y}$ to that of the section spanned by X, Y at p. First of all, note that the expression $[\bar{X}, \bar{Y}]^V|_p$ depends only on the values of $\bar{X}, \bar{Y}$ at p. In fact, if T is a vector field tangent to V, then

$$\begin{aligned}
\langle [\bar{X}, \bar{Y}], T \rangle &= \langle \bar{\nabla}_{\bar{X}} \bar{Y} - \bar{\nabla}_{\bar{Y}} \bar{X}, T \rangle \\
&= \bar{X} \langle \bar{Y}, T \rangle - \langle \bar{Y}, \bar{\nabla}_{\bar{X}} T \rangle - \bar{Y} \langle \bar{X}, T \rangle + \langle \bar{X}, \bar{\nabla}_{\bar{Y}} T \rangle \\
&= \langle \bar{X}, \bar{\nabla}_{\bar{Y}} T \rangle - \langle \bar{Y}, \bar{\nabla}_{\bar{X}} T \rangle.
\end{aligned}$$

THEOREM 3.20 (O'Neill). $K(X, Y) = K(\bar{X}, \bar{Y}) + \frac{3}{4} \| [\bar{X}, \bar{Y}]^V \|^2$.

Thus Riemannian submersions are curvature nondecreasing on the horizontal sections.

Let $\pi : M \to N$ be any smooth map. Vector fields $\bar{X}, X$ are called *π-related* if at all points $q \in \pi^{-1}(p)$,

$$\mathrm{d}\pi(\bar{X}) = X.$$

In particular $\bar{X}, X$ as above are π-related. We shall make use of the following lemma.

LEMMA 3.21. *If $\tilde{X}$, X and $\tilde{Y}$, Y are π-related, then $[\tilde{X}, \tilde{Y}]$ is π-related to $[X, Y]$.*

PROOF. For any function $f : N \to \mathbb{R}$,

$$\tilde{X}(f \circ \pi) = \mathrm{d}\pi(\tilde{X})(f) = X(f).$$

Therefore at $q \in \pi^{-1}(p)$,

$$\mathrm{d}\pi([\tilde{X}, \tilde{Y}])(f) = (\tilde{X}\tilde{Y} - \tilde{Y}\tilde{X})(f \circ \pi) = (XY - YX)f = [X, Y](f).$$

Since a tangent vector is determined by its action on functions, the lemma follows. □

PROOF OF THEOREM 3.20. Lemma 3.21, together with the Riemannian submersion property, has the consequence that given X, Y, Z on N and a vertical field T on M,

$$\langle[\bar{X},\bar{Y}],\bar{Z}\rangle = \langle[X,Y],Z\rangle, \qquad \langle[\bar{X},T],\bar{Y}\rangle = 0. \tag{3.22}$$

Then the formula for the Riemannian connection of Chapter 1, Section 1, together with (3.22) and the Riemannian submersion property, gives

$$\begin{aligned}\langle\bar{\nabla}_{\bar{X}}\bar{Y},\bar{Z}\rangle &= \frac{1}{2}\{\bar{X}\langle\bar{Y},\bar{Z}\rangle + \bar{Y}\langle\bar{X},\bar{Z}\rangle - \bar{Z}\langle\bar{X},\bar{Y}\rangle \\ &\quad + \langle[\bar{X},\bar{Y}],\bar{Z}\rangle - \langle[\bar{X},\bar{Z}],\bar{Y}\rangle - \langle[\bar{Y},\bar{Z}],\bar{X}\rangle\} \\ &= \langle\nabla_X Y, Z\rangle,\end{aligned} \tag{3.23}$$

while if T is vertical,

$$\begin{aligned}\langle\bar{\nabla}_{\bar{X}}\bar{Y},T\rangle = \frac{1}{2}\{&\bar{X}\langle\bar{Y},T\rangle + \bar{Y}\langle\bar{X},T\rangle - T\langle\bar{X},\bar{Y}\rangle \\ &+ \langle[\bar{X},\bar{Y}],T\rangle - \langle[\bar{X},T],\bar{Y}\rangle - \langle[\bar{Y},T],\bar{X}\rangle\}.\end{aligned}$$

Since $\langle\bar{X},\bar{Y}\rangle = \langle X,Y\rangle$ and T is vertical, $T\langle\bar{X},\bar{Y}\rangle = 0$. The first two terms on the right clearly vanish as do the last two, by (3.22). Therefore

$$\langle\bar{\nabla}_{\bar{X}}\bar{Y},T\rangle = \frac{1}{2}\langle[\bar{X},\bar{Y}],T\rangle. \tag{3.24}$$

Thus by (3.23) and (3.24),

$$\bar{\nabla}_{\bar{X}}\bar{Y} = (\overline{\nabla_X Y}) + \frac{1}{2}[\bar{X},\bar{Y}]^V. \tag{3.25}$$

Also, by (3.23) and (3.24),

$$\begin{aligned}\langle\nabla_T\bar{X},\bar{Y}\rangle &= \langle\bar{\nabla}_{\bar{X}}T,\bar{Y}\rangle + \langle[T,\bar{X}],\bar{Y}\rangle \\ &= -\,\langle\bar{\nabla}_{\bar{X}}\bar{Y},T\rangle = -\frac{1}{2}\langle[\bar{X},\bar{Y}],T\rangle = -\frac{1}{2}\langle[\bar{X},\bar{Y}]^V,T\rangle.\end{aligned} \tag{3.26}$$

Now by (3.23) it is clear that

$$\bar{X}\langle\bar{\nabla}_{\bar{Y}}\bar{Z},\bar{W}\rangle = X\langle\nabla_Y Z,W\rangle. \tag{3.27}$$

Therefore

$$\begin{aligned}\langle\bar{\nabla}_{\bar{X}}\bar{\nabla}_{\bar{Y}}\bar{Z},\bar{W}\rangle &= \bar{X}\langle\bar{\nabla}_{\bar{Y}}\bar{Z},\bar{W}\rangle - \langle\bar{\nabla}_{\bar{Y}}\bar{Z},\bar{\nabla}_{\bar{X}}\bar{W}\rangle \\ &= X\langle\nabla_Y Z,W\rangle - \langle\nabla_Y Z,\nabla_X W\rangle - \frac{1}{4}\langle[\bar{Y},\bar{Z}],[\bar{X},\bar{W}]\rangle \\ &= \langle\nabla_X\nabla_Y Z,W\rangle - \frac{1}{4}\langle[\bar{Y},\bar{Z}]^V,[\bar{X},\bar{W}]^V\rangle.\end{aligned} \tag{3.28}$$

Also by (3.23) and (3.26),

$$\begin{aligned}\langle\bar{\nabla}_{[\bar{X},\bar{Y}]}\bar{Z},\bar{W}\rangle &= \langle\bar{\nabla}_{[\bar{X},\bar{Y}]^H}\bar{Z},\bar{W}\rangle + \langle\bar{\nabla}_{[\bar{X},\bar{Y}]^V}\bar{Z},\bar{W}\rangle \\ &= \langle\nabla_{[X,Y]}Z,W\rangle - \frac{1}{2}\langle[\bar{Z},\bar{W}]^V,[\bar{X},\bar{Y}]^V\rangle.\end{aligned} \tag{3.29}$$

Therefore, using (3.28) and (3.29),

$$\begin{aligned}\langle \bar{R}(\bar{X},\bar{Y})\bar{Z},\bar{W}\rangle &= \langle \bar{\nabla}_{\bar{X}}\bar{\nabla}_{\bar{Y}}\bar{Z},\bar{W}\rangle - \langle \bar{\nabla}_{\bar{Y}}\bar{\nabla}_{\bar{X}}\bar{Z},\bar{W}\rangle - \langle \bar{\nabla}_{[\bar{X},\bar{Y}]}\bar{Z},\bar{W}\rangle \\ &= \langle R(X,Y)Z,W\rangle + \frac{1}{4}\langle [\bar{X},\bar{Z}]^V,[\bar{Y},\bar{W}]^V\rangle \\ &\quad - \frac{1}{4}\langle [\bar{Y},\bar{Z}]^V,[\bar{X},\bar{W}]^V\rangle + \frac{1}{2}\langle [\bar{Z},\bar{W}]^V,[\bar{X},\bar{Y}]^V\rangle.\end{aligned} \tag{3.30}$$

The theorem follows by setting $\bar{Z}=\bar{Y}$, $\bar{X}=\bar{W}$. □

We make one more general remark:

PROPOSITION 3.31. *If $\pi : M \to N$ is a Riemannian submersion, $\gamma : [0,1] \to N$ and $\bar{\gamma} : [0,1] \to M$ a horizontal lift, then γ is a geodesic if and only if $\bar{\gamma}$ is.*

PROOF. From (3.25),

$$\bar{\nabla}_{\bar{\gamma}'}\bar{\gamma}' = \overline{\nabla_{\gamma'}\gamma'} + \frac{1}{2}[\bar{\gamma}',\bar{\gamma}']^V = \overline{\nabla_{\gamma'}\gamma'},$$

and the claim follows immediately. □

Proposition 3.31 may be seen more geometrically from the relation

$$L[\phi] = \int ||\phi'||\mathrm{d}t \geq \int ||(\phi')^H||\mathrm{d}t = L[\pi(\phi)]$$

for any curve $\phi : [0,1] \to M$, and the relation

$$L[\bar{\gamma}] = L[\gamma].$$

We now specialize back to the case of homogeneous spaces. The map $\pi : G \to G/H$ is a fibration and hence a submersion. If G/H admits a left invariant metric $\langle\langle\ ,\ \rangle\rangle$ then by Proposition 3.16(4), G admits a left invariant metric $\langle\ ,\ \rangle$ which is right invariant under H. The restriction of $\langle\ ,\ \rangle$ to $\mathfrak{h}$ is bi-invariant, and its restriction to $\mathfrak{p}=\mathfrak{h}^\perp$ induces $\langle\langle\ ,\ \rangle\rangle$. Then $\pi : G \to G/H$ is a Riemannian submersion, and the curvature of G/H may be computed immediately from Proposition 3.18(3) and Theorem 3.20. The decomposition $\mathfrak{g} = \mathfrak{p}\bigoplus\mathfrak{h}$ corresponds precisely to the decomposition $M_q = H\bigoplus V$ and for $X, Y \in \mathfrak{p}$, the correction term in Theorem 3.20 becomes $\frac{3}{4}||[X,Y]_{\mathfrak{h}}||^2$. We get the formula

$$\begin{aligned}K(X,Y) &= ||(\mathrm{ad}_X)^*(Y) + (\mathrm{ad}_Y)^*(X)||^2 - \langle(\mathrm{ad}_X)^*(X),(\mathrm{ad}_Y)^*(Y)\rangle \\ &\quad - \frac{3}{4}||[X,Y]_{\mathfrak{p}}||^2 - \frac{1}{2}\langle[[X,Y],Y],X\rangle - \frac{1}{2}\langle[[Y,X],X],Y\rangle.\end{aligned} \tag{3.32}$$

Here we have written $K(X,Y)$ for $K(\mathrm{d}\pi(X),\mathrm{d}\pi(Y))$. We will continue to do this.

Suppose that the metric $\langle\ ,\ \rangle$ on G is bi-invariant. In this case the corresponding metric on G/H is called *normal*. As in Corollary 3.19, the formula for the curvature simplifies substantially.

COROLLARY 3.33. *If the metric on G/H is normal, then*
(1) $K(X,Y) = \frac{1}{4}\|[X,Y]_{\mathfrak{p}}\|^2 + \frac{1}{4}\|[X,Y]_{\mathfrak{h}}\|^2$.
In particular the sectional curvature is nonnegative.
(2) The geodesics in G/H are the images of one-parameter subgroups of G.

PROOF. (1) is immediate from Corollary 3.19(3) and Theorem 3.20. (2) Follows from Corollary 3.19(4) and Proposition 3.31. □

Berger [1961], has classified those normal homogeneous spaces which have *strictly positive* curvature. With two exceptions of dimension 7 and 13, they are precisely the symmetric spaces of rank 1 which will be introduced later in this chapter. Recently, Wallach [1972a,b] has discovered 3 new examples of (non-normal) homogeneous spaces of positive curvature of dimensions 6, 12, 24. He has shown that in even dimensions these are the only nonsymmetric homogeneous spaces of positive curvature. However, he has also constructed infinitely many simply connected nondiffeomorphic 7-dimensional homogeneous spaces with positive curvature.

PROPOSITION 3.34. *A simply connected Lie group which admits a bi-invariant metric is the product of a compact group and a vector group.*

PROOF. Let $\mathfrak{z}$ denote the center of $\mathfrak{g}$

$$\mathfrak{z} = \{x \in \mathfrak{g} | [x,y] = 0 \ \forall y \in \mathfrak{g}\}.$$

It is clear that $\mathfrak{z}$ is an ideal. On the other hand, if G admits a bi-invariant metric, and I_1 is any ideal of $\mathfrak{g}$, then $I_2 = I_1^{\perp}$ is also an ideal since

$$0 = \langle I_1, I_2 \rangle = \langle [x, I_1], I_2 \rangle = \langle I_1, [x, I_2] \rangle.$$

Therefore, in this case $\mathfrak{g}$ splits as $\mathfrak{g} = \mathfrak{z} \bigoplus \mathfrak{h}$. Let $G = Z \times H$ be the corresponding splitting of G. Then G is easily seen to be the isometric product of Z and H. Z is simply connected, abelian and therefore a vector group. The formula of Corollary 3.19 for the curvature of a Lie group with bi-invariant metric implies that if the corresponding Lie algebra has no center, then the Ricci curvature is strictly positive. Hence by Myers' Theorem 1.31, H is compact. □

Theorems 8.19 and 8.23 are much more general results of the type of Proposition 3.34. They are proved by quite different methods.

The following example is due to Berger. In addition to illustrating how our formulas work in practice, it provides a counterexample to a certain conjecture about closed geodesics. This point is explained in Chapter 5.

EXAMPLE 3.35 (Berger). Consider the 3-dimensional Lie algebra L that is spanned by elements z_1, z_2, z_3 with multiplication table

$$[z_1, z_2] = -2z_3, \qquad [z_1, z_3] = 2z_2, \qquad [z_2, z_3] = -2z_1.$$

It is straightforward to check that the inner product defined by making z_1, z_2, z_3 orthonormal is invariant under ad L. It is known that the simply connected Lie group with Lie algebra L is the ordinary 3-sphere S^3 which is a 2-fold covering space of $SO(3)$. In fact, by using the formula for the curvature, one may easily show that the sectional curvature is constant and equal to 1. We wish to consider the homogeneous space G/H with $G = S^3 + \mathbb{R}$, and H the one-parameter subgroup generated by $\alpha z_1 + \beta z_4$, where $z_1 \in L$, $\alpha^2 + \beta^2 = 1$ and z_4 is the l.i.v.f. tangent to $\mathbb{R}$. If $\beta \neq 0$, then G/H is easily seen to be diffeomorphic to S^3. In fact $\Pi|S^3 \times \{0\}$ is a non-singular smooth map from S^3 to G/H.

Further, $\Pi|S^3 \times \{0\}$ is trivially seen to be 1-1. (Notice that G does not act effectively on G/H). G/H may be given a normal metric by taking $||z_4|| = 1$ and $\langle z_i, z_4\rangle = 0$, $i = 1, 2, 3$. We note for future reference that $\gamma = \Pi\big(\exp(t(-\beta z_1 + \alpha z_4))\big)$ is a geodesic as follows from Corollary 3.33. In fact, γ is periodic of length $2\pi\beta$, since

$$\begin{aligned} e^{2\pi\beta(-\beta z_1+\alpha z_4)} &= e^{2\pi(-\beta^2 z_1+\beta\alpha z_4)} = e^{2\pi((\alpha^2-1)z_1+\beta\alpha z_4)} \\ &= e^{-2\pi z_1} e^{2\pi\alpha(\alpha z_1+\beta z_4)} = e^{2\pi\alpha(\alpha z_1+\beta z_4)}. \end{aligned}$$

($e^{-2\pi z_1} = e$ since e^{tz_1} is a periodic geodesic of length 2π in S^3.) Now if $z = \beta z_1 - \alpha z_4$, then

$$\begin{aligned} A &= \mu_1 z + \mu_2 z_2 + \mu_3 z_3, \\ B &= \nu_1 z + \nu_2 z_2 + \nu_3 z_3, \\ ([A, B]) &= 2\lambda_1 z_1 + 2\beta\lambda_2 z_2 + 2\beta\lambda_3 z_3, \\ \lambda_1 &= \mu_2\nu_3 - \mu_3\nu_2, \\ \lambda_2 &= \mu_1\nu_3 - \mu_3\nu_1, \\ \lambda_3 &= \mu_1\nu_2 - \mu_2\nu_1. \end{aligned}$$

Then one gets easily

$$\begin{aligned} ||[A, B]_{\mathfrak{h}}||^2 &= 4\alpha^2\lambda_1^2, \\ ||[A, B]||^2 &= 4\beta^2(\lambda_1^2 + \lambda_2^2 + \lambda_3^2), \\ ||A \wedge B||^2 &= \lambda_1^2 + \lambda_2^2 + \lambda_3^2 > 0, \\ K(A, B) &= \frac{(1 + 3\alpha^2)\lambda_1^2 + \beta^2(\lambda_2^2 + \lambda_3^2)}{\lambda_1^2 + \lambda_2^2 + \lambda_3^2}. \end{aligned}$$

One sees that the sectional curvature satisfies

$$H = \beta^2 \leq K_M \leq 1 + 3\alpha^2 = 4 - 3\beta^2 = K.$$

We note that if $\frac{\beta^2}{(4-3\beta^2)} < \frac{1}{9}$, then

$$2\pi\beta < 2\pi(4 - 3\beta^2)^{-\frac{1}{2}} = 2\pi K^{-\frac{1}{2}}$$

as an easy computation shows. In Chapter 5 we will show that for an even-dimensional manifold M with $K \geq K_M \geq H > 0$, every closed geodesic has length $\geq 2\pi K^{-\frac{1}{2}}$. If M is odd-dimensional with $K \geq K_M \geq \frac{1}{4}K$, the same result obtains. Our example, however, shows that such a result does not hold in general. The best one could hope for is $K \geq K_M \geq \frac{1}{9}K$.

We will now briefly discuss a special class of homogeneous spaces — the symmetric spaces.

DEFINITION 3.36. The Riemannian manifold M is called *locally symmetric* if for each $m \in M$ there exists r such that reflection through the origin (in normal coordinates) is an isometry on $B_r(m)$. M is *(globally) symmetric* if the above reflection extends to a global isometry $I_m : M \to M$.

PROPOSITION 3.37. *(1) M is locally symmetric if and only if $\nabla R = 0$.*

(2) M simply connected complete and locally symmetric implies M symmetric.

(3) M symmetric implies M homogeneous. $M = G/H$, where G is the isometry group of M and H is the isotropy group of some point $m \in M$.

(4) Let $M = G/H$ be symmetric with G the isometry group of M and H the isotropy group of $m \in M$. Let I denote the symmetry about m. Then $g \to IgI$ defines an automorphism σ of G such that $\sigma^2 = 1$. The set F of fixed points of σ is a closed subgroup containing H. Its identity component F_0 coincides with that of H.

(5) Conversely, let G be a Lie group, σ an automorphism such that $\sigma^2 = 1$, and $\langle\ ,\ \rangle$ a left invariant metric on G/F, where F is the set of fixed points of σ. The relation $\sigma(gf) = \sigma(g)\sigma(f) = \sigma(g)f$ shows that σ induces a diffeomorphism of G/F. If this diffeomorphism preserves $\langle\ ,\ \rangle$, then G/F is a symmetric space.[2]

(6) A simply connected Lie group G possesses an automorphism σ such that $\sigma^2 = 1$ if and only if $\mathfrak{g} = \mathfrak{p} \bigoplus \mathfrak{h}$ with

$$[\mathfrak{h},\mathfrak{h}] \subset \mathfrak{h}, \qquad [\mathfrak{h},\mathfrak{p}] \subset \mathfrak{p}, \qquad [\mathfrak{p},\mathfrak{p}] \subset \mathfrak{h}.$$

In case G/F admits a σ-invariant Riemannian metric, G/F is therefore globally symmetric and its curvature is given by

$$K(X,Y) = -\frac{1}{2}\langle[[X,Y],Y],X\rangle - \frac{1}{2}\langle[[Y,X],X],Y\rangle,$$

where $X, Y \in \mathfrak{p}$ are orthnormal.

PROOF. (1) Since I_m is an isometry, $\mathrm{d}I_m$ commutes with ∇R. Hence

$$\begin{aligned} -[\nabla_X R](y,z)w &= \mathrm{d}I_m([\nabla_X R](y,z)w) = [\nabla_{-X} R](-y,-z)(-w) \\ &= [\nabla_X R](y,z)w. \end{aligned}$$

The converse, which is an easy consequence of Lemma 1.41, is left as an exercise (or see Helgason [1962]).

[2]It can be shown that even if G/F does not admit a σ-invariant metric it nonetheless admits a unique left invariant affine connection preserved by σ. Thus in general G/F is *affine symmetric*.

(2) This follows immediately from Cartan-Ambrose-Hicks Theorem, using the condition $\nabla R = 0$ as above.

(3) Given a geodesic segment $\gamma : [0,t] \to M$,

$$\gamma(0,t) \cup I_{\gamma(t)}(\gamma(0,t)) \cup \dots$$

is γ extended arbitrarily far. Hence M is complete by the Hopf-Rinow Theorem 1.10. Given p, q, let $\gamma : [0, t_0] \to M$ be a geodesic segment from p to q. Then $I_{\gamma(\frac{t_0}{2})}(p) = q$. Hence M has a transitive group of isometries and is homogeneous.

(4) The only nontrivial part is to show $H_0 = F_0$. Since $H \subset F$ it suffices to check that $F_0 \subset H$. For $f \in F$, we have $f(m) = I \circ f \circ I(m) = I \circ f(m)$. However, m is the only point of a normal coordinate ball $B_r(m)$ which is fixed by I. Hence there is a neighborhood U of e in G such that $F \cap U \subset H$. It follows that $F \cap H$ is open (since $H \subset F$). But $F \cap H$ is also closed since F and H are closed. Hence $F_0 = H_0$.

(5) The symmetry about a coset $[gF]$ is given by $L_g \circ \sigma \circ L_{g^{-1}}$.

(6) If $\mathfrak{g} = \mathfrak{p} \bigoplus \mathfrak{h}$ as above, then the linear map $d\sigma$ defined by

$$d\sigma|\mathfrak{h} = 1, \qquad d\sigma|\mathfrak{p} = -1$$

is easily seen to be an automorphism of $\mathfrak{g}$ such that $(d\sigma)^2 = 1$. It induces an automorphism σ of G by the remarks following Proposition 3.4. Conversely given such a σ, take $\mathfrak{h}, \mathfrak{p}$ to be its $+1$ and -1 eigenspaces, respectively. Then for example

$$d\sigma([p_1, p_2]) = [d\sigma(p_1), d\sigma(p_2)] = [-p_1, -p_2] = [p_1, p_2]$$

shows $[\mathfrak{p}, \mathfrak{p}] \subset \mathfrak{h}$, and the other relations follow similarly.

If $X, Y, Z \in \mathfrak{p}$, then $[X, Z] \in \mathfrak{h}$ and

$$\langle (\mathrm{ad}_X)^*(Y), Z\rangle = \langle Y, [X, Z]\rangle = 0.$$

Therefore $(\mathrm{ad}_X)^* Y \subset \mathfrak{h}$. Now if $T \in \mathfrak{h}$,

$$\langle Y, [T, X]\rangle = -\langle [T, Y], X\rangle$$

because the metric on G is right invariant under H. Therefore

$$\langle (\mathrm{ad}_X)^*(Y), T\rangle = \langle Y, [X, T]\rangle = -\langle [Y, T], X\rangle = -\langle (\mathrm{ad}_Y)^*(X), T\rangle.$$

Therefore, substituting in (3.32) gives

$$K(X, Y) = -\frac{1}{2}\langle [[X, Y], Y], X\rangle - \frac{1}{2}\langle [[Y, X], X], Y\rangle.$$

□

A complete classification of symmetric spaces is available (Helgason [1962]). In particular, the only simply connected symmetric spaces having positive curvature are the spheres of constant curvature, complex and quaternionic projective spaces and the Cayley Plane. These are sometimes referred to as rank one symmetric spaces, and except for the spheres they have canonical metrics with sectional curvature varying between $\frac{1}{4}$ and 1.

As an example, we will compute the curvature of complex projective space. The calculations for the other rank one spaces are similar.

EXAMPLE 3.38. The unitary group $U(n)$ is defined as the (compact) group of $n \times n$ matrices with complex entries (α_{ij}) such that $(\alpha_{ij})^{-1} = (\bar{\alpha}_{ji})$. The special unitary group $SU(n)$ is the subgroup of $U(n)$ of matrices of determinant 1. The Lie algebra $\mathfrak{u}(n+1)$ of $U(n+1)$ consists of skew hermitian matrices $(\alpha_{ij}) = (-\bar{\alpha}_{ji})$. For $\mathfrak{su}(n+1)$ we must add the condition

$$\mathrm{trace}(\alpha_{ij}) = 0.$$

Complex projective space is the homogeneous space $CP(n) = SU(n+1)/U(n)$, where $U(n)$ is embedded in $SU(n+1)$ as

$$\left[\begin{array}{c|c} U & \mathbf{0} \\ \hline \mathbf{0} & \overline{\det U} \end{array}\right],$$

where $U \in U(n)$. Geometrically, $CP(n)$ may be thought of as the collection of 1-dimensional complex subspaces of $\mathbb{C}^{n+1}$.

The following may be easily checked. The rule $\langle A, B\rangle = -\frac{1}{2}\mathrm{trace}(AB)$ defines a bi-invariant metric on $SU(n+1)$ which gives rise to the decomposition $\mathfrak{su}(n+1) = \mathfrak{p} + \mathfrak{u}(n)$, where $\mathfrak{p}$ consists of matrices of the form

$$\alpha = \left[\begin{array}{c|c} & \bar{\alpha}_1 \\ \mathbf{0} & \vdots \\ & \bar{\alpha}_n \\ \hline \alpha_1 \ldots \alpha_n & \mathbf{0} \end{array}\right].$$

$\mathfrak{p}$ may be thought of as a complex n-space or real $2n$-space. Multiplication by i gives a real linear transformation $J : \mathfrak{p} \to \mathfrak{p}$ such that $J^2 = -1$ and $\langle x, y\rangle = \langle J(x), J(y)\rangle$. $[\mathfrak{p}, \mathfrak{p}] \subset \mathfrak{u}(n)$, so that $CP(n)$ is a symmetric space.

Now if $||\alpha|| = ||\beta|| = 1$ and $\langle \alpha, \beta\rangle = 0$, by Corollary 3.33,

$$\langle R(\alpha, \beta)\beta, \alpha\rangle = ||[\alpha, \beta]||^2,$$

where

$$[\alpha, \beta] = \left[\begin{array}{c|c} \beta_i\bar{\alpha}_j - \alpha_i\bar{\beta}_j & 0 \\ \hline 0 & \sum\limits_i \alpha_i\bar{\beta}_i - \beta_i\bar{\alpha}_i \end{array}\right].$$

$$
\begin{aligned}
||[\alpha,\beta]||^2 &= -\frac{1}{8}\sum_{ij}(\beta_i\bar{\alpha}_j - \alpha_i\bar{\beta}_j)(\beta_j\bar{\alpha}_i - \alpha_j\bar{\beta}_i) \\
&\quad -\frac{1}{8}\sum_{ij}(\alpha_i\bar{\beta}_i - \beta_i\bar{\alpha}_i)(\alpha_j\bar{\beta}_j - \beta_j\bar{\alpha}_j) \\
&= \sum_i \alpha_i\bar{\alpha}_i \sum_j \beta_j\bar{\beta}_j - \frac{1}{2}\sum_i \beta_i\bar{\alpha}_i \sum_j \beta_j\bar{\alpha}_j \\
&\quad -\frac{1}{2}\sum_i \alpha_i\bar{\beta}_i \sum_j \alpha_j\bar{\beta}_j \\
&\quad -\frac{1}{2}\sum_i(\bar{\beta}_i\alpha_i - \bar{\alpha}_i\beta_i)\sum_j(\alpha_j\bar{\beta}_j - \beta_j\bar{\alpha}_j) \\
&= \frac{1}{4}(||\alpha||^2||\beta||^2) + \frac{3}{4}\langle J(\alpha),\beta\rangle^2 = \frac{1}{4} + \frac{3}{4}\langle J(\alpha),\beta\rangle^2,
\end{aligned}
$$

where the last step follows from the relations

$$
\begin{aligned}
\sum(\alpha_i\bar{\beta}_i + \beta_i\bar{\alpha}_i) &= \langle \alpha,\beta\rangle, \\
\sum(-\alpha_i\bar{\beta}_i + \beta_i\bar{\alpha}_i) &= i\langle J(\alpha),\beta\rangle.
\end{aligned}
$$

If $\langle J(\alpha),\beta\rangle = 1$, then $K(\alpha,\beta) = 1$, while if $\langle J(\alpha),\beta\rangle = 0$, $K(\alpha,\beta) = \frac{1}{4}$.

Given a Lie algebra $\mathfrak{g}$, we define the *Killing form* as the form

$$B(g_1, g_2) = \operatorname{trace}(\operatorname{ad}_{g_2}, \operatorname{ad}_{g_1}).$$

Then B is easily seen to be symmetric. Further, for all $x \in \mathfrak{g}$, ad_x is skew symmetric with respect to B; i.e.

$$B(\operatorname{ad}_x y_1, y_2) = -B(y_1, \operatorname{ad}_x y_2).$$

If B is nondegenerate, we say that $\mathfrak{g}$ is *semi-simple.* It is easy to show that this implies that $\mathfrak{g}$ is a direct sum of *simple ideals.* An ideal is called *simple* if it contains no proper ideal. It is also true that if $\mathfrak{g}$ is a direct sum of simple ideals, then $\mathfrak{g}$ is semi-simple; but this is more difficult to prove; see Helgason [1962].

PROPOSITION 3.39. *(1) If the group G is compact, the Killing form B of $\mathfrak{g}$ is negative semi-definite. If B is negative definite, G is compact.*

(2) If G is semi-simple and noncompact, then $\mathfrak{g}$ may be decomposed as $\mathfrak{p} \bigoplus \mathfrak{h}$ such that $B|\mathfrak{h}$ is negative definite, $B|\mathfrak{p}$ is positive definite, $\mathfrak{h}$ is a (maximal compact) subalgebra, $[\mathfrak{h},\mathfrak{h}] \subset \mathfrak{h}$, $[\mathfrak{h},\mathfrak{p}] \subset \mathfrak{p}$, $[\mathfrak{p},\mathfrak{p}] \subset \mathfrak{h}$.

(3) Let G be semi-simple and non-compact, and let $\mathfrak{g} = \mathfrak{p} \bigoplus \mathfrak{h}$ be as above. Then $B|\mathfrak{p}$ defines a metric invariant under $\mathrm{d}\sigma$ *so that G/H becomes a symmetric space. The curvature of G/H is given by*

$$K(X,Y) = -||[X,Y]||^2.$$

(4) Let G be compact semi-simple. $\sigma^2 = I$, and $\mathfrak{g} = \mathfrak{p} \bigoplus \mathfrak{h}$ the decomposition into the + and − eigenspaces of σ. Then $-B|\mathfrak{g}$ is a bi-invariant metric and defines a symmetric metric on G/H. The curvature is given by $K(X,Y) = ||[X,Y]||^2$.

(5) Let $\mathfrak{g} = \mathfrak{p} \bigoplus \mathfrak{h}$ be as in (3) and let $[\ ,\ ]$ denote the Lie algebra structure. Then the rule

$$(3.40)\qquad [\![p_1, p_2]\!] = -[p_1, p_2], \qquad [\![p, h]\!] = [p, h], \qquad [\![h_1, h_2]\!] = [h_1, h_2]$$

defines a new Lie algebra structure $[\![\ ,\ ]\!]$ on $\mathfrak{g}$. Let $\hat{B}$ denote the Killing form of $[\![\ ,\ ]\!]$. Then

$$\hat{B}|\mathfrak{p} = -B|\mathfrak{p}, \qquad \hat{B}(\mathfrak{p},\mathfrak{h}) = 0, hatB|\mathfrak{h} = B|\mathfrak{h}.$$

In particular, B is negative definite.

Conversely, given $[\![\ ,\ ]\!]$, define $[\ ,\ ]$ by (3.40). Then $[\ ,\ ]$, $\mathfrak{g} = \mathfrak{p} \bigoplus \mathfrak{h}$ are as in (2).

(6) If G is a compact Lie group, then $G = (G \times G)/G$ is a symmetric space, where $G \subset G \times G$ is the diagonal inclusion.

PROOF. (1) If G is compact, then it admits a bi-invariant metric $\langle\ ,\ \rangle$. For all g, Ad_g is in the orthogonal group of $\langle\ ,\ \rangle$. Therefore, for all x, ad_x is represented by a skew symmetric matrix. But the trace of the square of a non-zero skew symmetric matrix is negative. If, on the other hand, B is negative definite, then $-B$ induces a bi-invariant metric on G. Using the fact that $\mathfrak{g}$ has no center, the claim follows from Proposition 3.34.

(2) See Helgason [1962], p. 156.

(3) Everything but the formula for the curvature follows immediately from (2). Moreover, if $X, Y \in \mathfrak{p}$, then $[[X,Y],Y] \in \mathfrak{p}$, and by Proposition 3.37(6),

$$\begin{aligned} K(X,Y) &= -\frac{1}{2}\langle[[X,Y],Y],X\rangle - \frac{1}{2}\langle[[Y,X],X],Y\rangle \\ &= -\frac{1}{2}B([[X,Y],Y],X) - \frac{1}{2}B([[Y,X],X],Y) \\ &= +\frac{1}{2}B([X,Y],[X,Y]) + \frac{1}{2}B([Y,X],[Y,X]). \end{aligned}$$

But $[X,Y] \in \mathfrak{h}$ and $B|\mathfrak{h} = -\langle\ ,\ \rangle|\mathfrak{h}$. Therefore the above expression becomes $-||[X,Y]||^2$.

(4) This follows by an argument completely analogous to that of (3).

(5) This is straightforward, and we omit the details.

(6) Set $\sigma(g_1, g_2) = (g_2, g_1)$. □

The metric in the Berger example is, of course, $-B$. The symmetry about e in G considered as a symmetric space is given by $g \to g^{-1}$.

A pair of symmetric spaces related as in (5) are called *dual* to one another.

We note that in a locally symmetric space since $\nabla R = 0$, the Jacobi equation $\nabla_T \nabla_T J = R(T,J)T$ has constant coefficients and hence may be solved explicitly. In fact, if $T, E_1, \ldots, E_n$ are an orthonormal base of eigenvectors of $x \to R(T,x)T$, at $t = 0$, then solutions vanishing at $t = 0$ are of the form

$$\sin(t\sqrt{\lambda})E(t), \qquad \sinh(t\sqrt{-\lambda})E(t), \qquad tE(t)$$

according as $\lambda > 0, \lambda < 0, \lambda = 0$, where $E(t)$ is a parallel eigenvector with eigenvalue λ. In particular, if G is a compact Lie group then

$$R(T,J)T = -\frac{1}{4}(\mathrm{ad}T)^2(J).$$

Since the nonzero eigenvalues of the square of a skew symmetric matrix occur in pairs, we have:

PROPOSITION 3.41 (Bott). *If G is a compact Lie group with bi-invariant metric, then all conjugate points are of even order.*

The following result will be utilized in Chapter 5 in the proof that in a symmetric space geodesics minimize up to the first conjugate point. Suppose that G is compact, $\sigma : G \to G$, $\sigma^2 = 1$ and H is the fixed point set of σ. Let

$$\mathfrak{g} = \mathfrak{p} \oplus \mathfrak{h}, \qquad d\sigma(\mathfrak{h}) = \mathfrak{h}, \qquad d\sigma(\mathfrak{p}) = -\mathfrak{p}.$$

Call $g \in G$ a *transvection* if $g \in \exp \mathfrak{p} = \mathfrak{L}$. $g \in \mathfrak{L}$ clearly implies $\sigma(g) = g^{-1}$, but not conversely.

PROPOSITION 3.42. *The map $\Phi : [gH] \to g\sigma(g^{-1})$ defines an embedding of G/H onto $\mathfrak{L}$. If $\langle\ ,\ \rangle$ is a bi-invariant metric, then $\mathfrak{L}$ is a totally geodesic submanifold, and the metric on G/H induced by Φ is 2 times the normal metric.*

PROOF. (1) Φ is well defined:

$$\begin{aligned} gh\sigma((gh)^{-1}) &= gh\sigma(h^{-1}g^{-1}) = gh\sigma(h^{-1})\sigma(g^{-1}) = ghh^{-1}\sigma(g^{-1}) \\ &= g\sigma(g^{-1}). \end{aligned}$$

(2) Φ is injective:

$$g\sigma(g^{-1}) = f\sigma(f^{-1})$$

implies

$$f^{-1}g = \sigma(f^{-1})(\sigma(g^{-1}))^{-1} = \sigma(f^{-1})\sigma(g) = \sigma(f^{-1}g).$$

This implies $f^{-1}g \in H$ or $[fH] = [gH]$.

(3) $\Phi(G/H) = \mathfrak{L}$: G/H is compact and therefore complete. Hence $\exp_{[H]}|G/H_{[H]}$ is onto. $d\pi : \mathfrak{p} \to G/H_{[H]}$ is injective. By Corollary 3.33, one-parameter subgroups of G are geodesics and project to geodesics in G/H. Therefore $\pi \circ \exp$ maps $\mathfrak{p}$ onto G/H. Given g and let $e^p \in \mathfrak{L}$ be such that $[gH] = [e^pH]$. Then

$$\Phi(g) = \Phi(e^p) = e^p\sigma(e^{-p}) = e^{2p} \in \mathfrak{L}$$

Thus $\Phi(G/H) \subset \mathfrak{L}$. But given $g = e^p \in \mathfrak{L}$,

$$\Phi(e^{p/2}) = e^{p/2}\Phi(e^{-p/2}) = e^p = g.$$

(4) Let $B_r(e)$ be a normal coordinate ball. Then

$$B_r(e) \cap \mathfrak{L} = \exp_e(B_r(0) \cap \mathfrak{p}).$$

Clearly $\exp_e(B_r(0) \cap \mathfrak{p}) \subset B_r(e) \cap \mathfrak{L}$.

Conversely if $g \in B_r(e) \cap \mathfrak{L}$, then σ maps the unique minimal geodesic

$$e^{tx} = \exp tx : [0, 1] \to M$$

from e to g into the unique minimal geodesic from e to $g^{-1} = \exp(-x) = e^{-x}$. But this geodesic is just e^{-tx}, so that $g = e^x$ with $\mathrm{d}\sigma(x) = -x$.

(5) $\Phi \circ \pi(g) = g^2$ if $g \in \mathfrak{L}$. Thus $\mathrm{d}\Phi \circ \mathrm{d}\pi|\mathfrak{p} = 2I$, where I is the identity map. In fact,

$$\Phi \circ \pi(g) = g\sigma(g^{-1}) = gg = g^2.$$

(6) Φ is an embedding and the metric on G/H induced by Φ is 2 times the normal metric: At $[H]$ this follows from (5), so it suffices to prove the relation

$$e^p\Phi([e^{-p}g])e^p = \Phi(g)$$

for all $e^p \in \mathfrak{L}$. For then, taking $[e^pH] = [gH]$ as is possible by (5), the above implies

$$\mathrm{d}R_g \circ \mathrm{d}L_g \circ \mathrm{d}\Phi_e = \mathrm{d}\Phi_g.$$

But, in fact,

$$e^p\Phi([e^{-p}g])e^p = e^pe^{-p}g\sigma(g^{-1}e^p)e^p = g\sigma(g^{-1})e^{-p}e^p = \Phi(g).$$

(7) $\mathfrak{L}$ is a totally geodesic submanifold: By (3),(4) it follows that if $B_r(e)$ is a normal coordinate ball then $\mathfrak{L} \cap B_r(e)$ is a submanifold which is geodesic at e. Therefore it suffices to verify that $e^{p/2}\mathfrak{L}e^{p/2} = \mathfrak{L}$. But if $e^x \in \mathfrak{L}$, then by (3),

$$e^{p/2}e^xe^{p/2} = \Phi(e^{p/2}e^{x/2}) \in \mathfrak{L}.$$

This completes the proof. □

We shall now briefly describe some applications of Theorem 3.20 to the problem of finding examples of manifolds which admit a metric of nonnegative curvature but are not diffeomorphic to a homogeneous space.

EXAMPLE 3.43. Let G be a Lie group, K a compact subgroup and $\langle\ ,\ \rangle$ a metric right invariant under K. Let M be a manifold on which K acts by isometries. Then K acts by isometries on the product $G \times M$ by $k(g, m) = (gk^{-1}, km)$. Clearly K acts without fixed point, and the quotient, which we write as $G \times_K M$, is a manifold.

$\Pi : G \times M \to G \times_K M$ is a submersion. Topologically, $G \times_K M$ is the bundle with fiber M, associated with the principal fibration $K \to G \to G/K$. Since K acts by isometries, $G \times_K M$ naturally inherits a metric such that Π becomes a Riemannian submersion. If the metrics on G and M have nonnegative curvature, then by Theorem 3.20 so does the metric on $G \times_K M$. For G compact, we could take the bi-invariant metric on G, which does have nonnegative curvature. Even if M is homogeneous, if K does not act transitively then $G\times_K M$ is not, in general, homogeneous; see Cheeger [1973] for details.

EXAMPLE 3.44 (Gromoll and Meyer). Let $Sp(n)$ denote the n-dimensional symplectic group realized by $n \times n$ matrices Q, with quaternionic entries, satisfying $Q\bar{Q}^t = I$. $Sp(n)$ is a compact Lie group and carries a bi-invariant metric. Let $Sp(1)$ act on $Sp(2)$ by

$$Q \to \begin{bmatrix} q & 0 \\ 0 & q \end{bmatrix} Q \begin{bmatrix} q & 0 \\ 0 & 1 \end{bmatrix}, \qquad q \in Sp(1), \qquad Q \in Sp(2).$$

This action is free and the quotient space turns out to be an exotic 7-sphere. Thus this exotic sphere carries a metric of nonnegative curvature. A calculation shows that on an open dense set of points the curvature is actually strictly positive.

At present all known examples of manifolds of nonnegative curvature are constructed by techniques closely related to the above.

CHAPTER 4

Morse Theory

Morse Theory is the study of the topology of the differentiable manifolds by means of the analysis of critical-point behavior of smooth functions. Specifically, if one is given a manifold M and a generic function $f : M \to \mathbb{R}$, one can find a homotopy equivalence between M and a CW-complex which has one cell for each critical point of f.

This theory was originated by Marston Morse (ca. 1930; see Morse [1934]) and has been a very fruitful tool in the study of differential geometry and topology. Since a beautiful exposition (Milnor [1963]) is already available, we will merely discuss those results which will be needed in other chapters, and for the most part rely on Milnor for the proofs. We begin with the study of critical points.

DEFINITION 4.1. Let $f : M \to \mathbb{R}$ and $p \in M$. Then p is called a *critical point* of f if $(\mathrm{d}f)_p = 0$; i.e. $(\mathrm{d}f)_p$ is the zero element of M_p^*. We call c a *critical value* of f if there exists some critical point p of f such that $f(p) = c$.

LEMMA 4.2. *If p is a critical point of $f : M \to \mathbb{R}$, and X and Y are vector fields on M, then $(X(Yf))(p)$ depends only on $X(p)$ and $Y(p)$. Furthermore, the map $X(p), Y(p) \to X(Y(f))(p)$ is a symmetric bilinear form on M_p.*

PROOF. We have

$$(X(Yf))(p) - (Y(Xf))(p) = ([X,Y])f(p) = \mathrm{d}f([X,Y])_p = 0,$$

so $X(Yf)(p)$ is symmetric in the vector fields X and Y. Also $X(Yf)(p)$ depends only on $X(p)$ and $Y(Xf)(p)$ depends only on $Y(p)$, so $X(Yf)(p)$ depends only om $X(p), Y(p)$ and is symmetric by the above. It is clearly bilinear, so the lemma follows. □

DEFINITION 4.3. Let p be a critical point of f, and let $H_pf : M_p \times M_p \to \mathbb{R}$ be the map defined by the above lemma. H_pf is called the *Hessian* of f. If H_pf is a nondegenerate bilinear form, i.e. if the map $x \to H_pf(x,)$ is an isomorphism of M_p onto M_p^*, we say that p is a *nondegenerate critical point*. The dimension λ of the subspace of maximum dimension on which H_pf is negative definite is called the *index* of p.

Let f be expressed in local coordinates with the origin at p. Let $D_x^2 f$ denote the matrix whose i, j-th entry is $\partial^2 f(x)/\partial x_i \partial x_j$. Via the standard inner product we may identify symmetric matrices with symmetric bilinear forms. With this identification it is easy to see that $H_pf = D_0^2 f$.

All manifolds admit functions which have only nondegenerate critical points. In fact, any function can be approximated in the $\mathbf{C}^r$ topology by a function of this type; see Milnor [1963], p. 32.

We now state the Morse Lemma, which gives a canonical form for f in a neighboring of any nondegenerate critical point. Our proof is from some notes of J.J. Duistermaat.

LEMMA 4.4. *Let p be a nondegenerate critical point of $f : M \to \mathbb{R}$. Then there exists about p a chart $\phi : U \to M$*

$$f \circ \phi(x) = f(p) + x_1^2 + \cdots + x_{n-\lambda}^2 - y_1^2 - \cdots - y_\lambda^2.$$

To prove Lemma 4.4, we shall need some elementary notions about bilinear maps.

Let V be a finite-dimensional vector space, and let $Q : V \times V \to \mathbb{R}$ be a bilinear map. Then if V^* is the dual space of V, $\tilde{Q} : V \to V^*$ is defined by the equation

$$\tilde{Q}(v)(w) = Q(v, w).$$

Let $L(V)$ be the space of linear transformations from V to itself, and let $SB(V)$ be the symmetric bilinear forms on V, i.e., $Q(v,w) = Q(w,v)$. Given $Q \in SB(V)$, define $G : L(V) \to SB(V)$ by

$$G(S)(v,w) = Q(S(v), S(w)).$$

LEMMA 4.5. *If Q is nondegenerate, $\mathrm{d}G$ has maximal rank in a neighborhood of I, where I is the identity map in $L(V)$. Therefore there exists a neighborhood N of Q in $SB(V)$ and a smooth map $J : SB(V) \to L(V)$ such that $G \circ J$ is the identity on N.*

PROOF. The second statement follows immediately from the first by the implicit-function theorem. To prove the first statement we need only show that at I, $\mathrm{d}G : L(V) \to SB(V)$ is surjective. But

$$\mathrm{d}G(S)(v,w) = Q(S(v), w) + Q(v, S(w)). \qquad (*)$$

Given $C \in SB(V)$, we can check by using $(*)$ that at I, $\mathrm{d}G(S) = C$ if $S = \frac{1}{2}\tilde{Q}^{-1}C$. The lemma follows. □

Proof of Lemma 4.4. Fix a chart $\psi : \mathcal{O} \to M$ about p, where $\mathcal{O}$ is a neighborhood of the origin in $\mathbb{R}^n$ and $\psi(0) = p$. Then it suffices to prove the lemma for $g = f \circ \psi : \mathcal{O} \to \mathbb{R}$, which has a critical point at the origin. We can assume that $H_0 g = D_0^2 g$ is in diagonal form with $n - \lambda$ eigenvalues equal to 1 and λ eigenvalues equal to -1.

Denote by B_z the element of $SB(R^n)$ defined by

$$B_z(v,w) = \int_0^1 D_{tz}^2 g(v,w)\mathrm{d}t.$$

Then $B_0 = H_0 g$ and, since the origin is a critical point, by Taylor's Theorem we have

$$g(z) = g(0) + B_z(z,z).$$

Let $Q = H_0g \in SB(\mathbb{R}^n)$, and let G, J be as in Lemma 4.5. Then for sufficiently small z, B_z is in the domain of J. For such z we may define $\eta(z)$ by

$$\eta(z) = J(B_z)(z).$$

Then

$$B_z(z,z) = H_0g(\eta(z), \eta(z)).$$

Note that at the origin, $d\eta = I$. Therefore, by the implicit function theorem, η is a diffeomorphism; $\eta : U_0 \to U$, for suitable neighborhoods U, U_0 of the origin. Set $\phi = \eta^{-1}$. Then

$$g(\phi(x)) = g(0) + B_{\phi(x)}(\phi(x), \phi(x)) = g(0) + H_0g(x,x).$$

Since H_0g is in diagonal form, in coordinates this just says

$$g(\phi(x)) = g(0) + x_1^2 + \cdots + x_{n-\lambda}^2 - y_1^2 - \cdots - y_\lambda^2.$$

□

COROLLARY 4.6. *Non degenerate critical points are isolated.*

PROOF. It is clear from the form of $f \circ \phi : U \to \mathbb{R}$ that its only critical point is at 0. Thus f has no critical points in U except p, so p is isolated. □

Now we begin to study the relationship between critical points and homotopy type. Let $f : M \to \mathbb{R}$ and let

$$M^a = \{p \in M | f(p) \leq a\}.$$

LEMMA 4.7. *If a is not a critical value of f, then M^a is a manifold with boundary $f^{-1}(a)$.*

PROOF. Given $p \in M^a$, $f(p) < a$, then there is a neighborhood of p in M which is included in M^a. If $f(p) = a$, and p is not a critical point of a, then by the implicit-function theorem one can find functions, $x_2, \ldots, x_n$ which together with f form a coordinate system near p. One then gets a half-space chart for M^a by the restriction $f(x) \leq a$. □

THEOREM 4.8. *If $f^{-1}[a,b]$ is compact and contains no critical points of f, then M^a is diffeomorphic to M^b. In fact M^a is a deformation retract of M^b, which is a diffeomorphism at each stage. Therefore the inclusion $M^a \to M^b$ is a homotopy equivalence.*

PROOF. See Milnor [1963], p. 12. The idea is to fix a Riemannian structure on M and consider the gradient vector field grad$f$ of f. Then M^b is deformed into M^a by pushing it at constant speed down the integral curves of grad$f$. The fact that grad$f$ is nowhere zero in $f^{-1}[a,b]$ is crucial, as is the compactness of $f^{-1}[a,b]$. (See Fig. 4.1). □

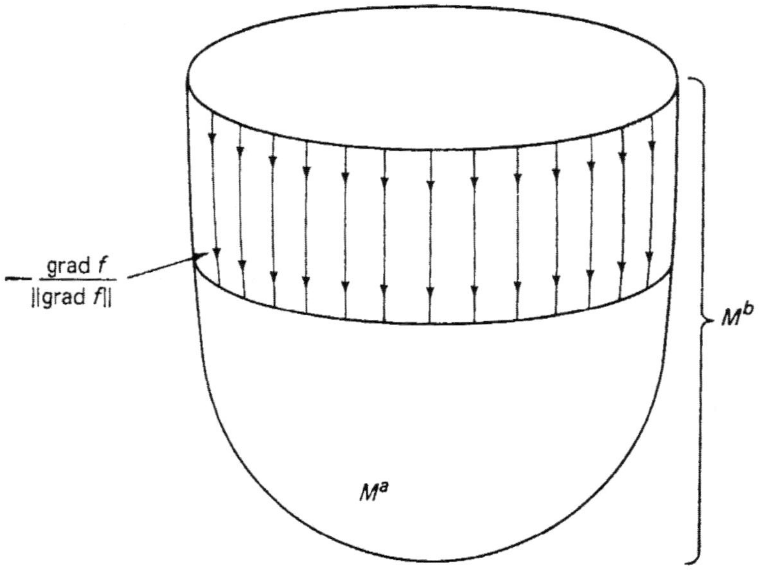

Fig. 4.1.

THEOREM 4.9. *Let $f : M \to \mathbb{R}$ have a nondegenerate critical point p of index λ such that $f(p) = c$. Assume that there exists ϵ such that $f^{-1}[c-\epsilon, c+\epsilon]$ is compact and contains no critical points except p. Then $M^{c+\epsilon}$ has the homotopy type of $M^{c-\epsilon}$ with a cell of dimension λ attached. (See Fig. 4.2).*

PROOF. See Milnor [1963], p. 14. □

THEOREM 4.10. *If $f : M \to \mathbb{R}$ has only nondegenerate critical points, and for all $a \in \mathbb{R}$, M^a is compact, then M has the homotopy type of a CW-complex with one cell of dimension λ for each critical point of index λ.*

PROOF. See Milnor [1963], p. 20. This theorem, of course, follows from repeated applications of Theorem 4.9. □

Note that any function satisfying the hypothesis of the above theorem must be bounded below. For if $\{p_i\}_{i=1}^{\infty}$ is a sequence of points such that $f(p_i) \to -\infty$, then $\{p_i\} \subseteq M^a$ for some a and $\{p_i\}$ clearly has no limit point and hence no convergent subsequence.

The following lemma will prove useful in the next chapter.

LEMMA 4.11. *(1) Let $f : M \to \mathbb{R}$ be as in the above theorem, except that critical points may be degenerate. Let $p, q \in M$ and let $\gamma : [0,1] \to M$ be a smooth curve from p to q. Let $a = \max\{f(p), f(q)\}$. Then given $\epsilon > 0$, γ is homotopic, keeping endpoints fixed to a curve $\hat{\gamma}$, where $f|\hat{\gamma} \le \max(a, c) + \epsilon$ and c is the largest critical value of index one or zero.*

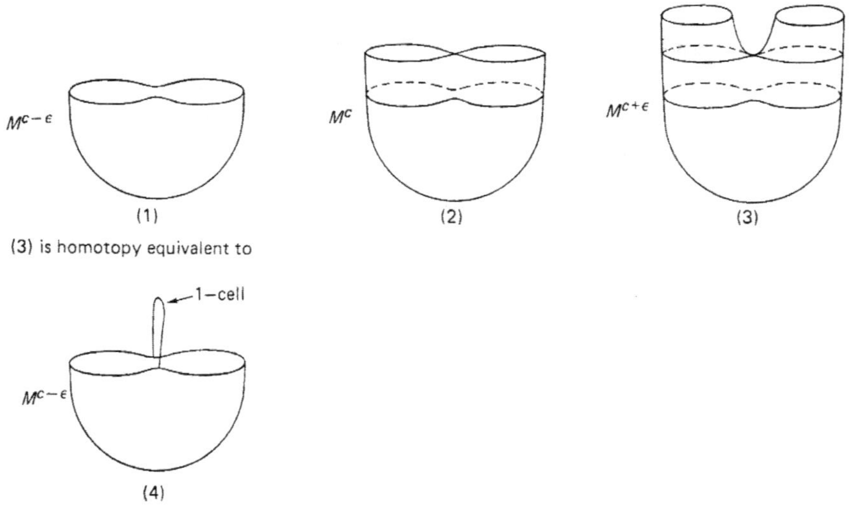

Fig. 4.2.

(2) If M is connected, f has no degenerate critical points, and if there are no critical points of index one, then there is at most one critical point of index zero.

PROOF. (1) First assume that f has nondegenerate critical points. Let $b = \max\{f|\gamma[0,1]\}$, and let $p_1, \ldots, p_k$ be the critical points of f in $M^b - M^a$ with values $f(p_i) = c_i$. Since the critical points are isolated, it is easy to alter f (without changing its critical set and without changing a and b) so that $\{c_i\}$ are distinct. So assume $b \geq c_1 > c_2 > \cdots > c_k > a$.

For any ϵ, there exists a deformation retract $h_s^1 : M^b \to M^{c_1+\epsilon}$ by Theorem 4.8. Let $\gamma_1 = h_1^1\gamma$. If p_1 has index zero, it is a strict local minimum, and if ϵ is small enough, p is contained in a component H of $M^{c_1+\epsilon}$ such that $f(H) \subseteq [c_1, c_1+\epsilon]$. Therefore, either $\gamma_1 \subseteq H$ or $\gamma_1 \cap H = \emptyset$.

If $\gamma_1 \subseteq H$, then $f(p), f(q) \geq c_1$, which contradicts $c_1 > a$. Thus $\gamma_1 \cap H = \emptyset$, so if p_1 has index zero we can forget about it. Hence we assume that p_1 has index at least two.

From the proof of Theorem 4.9 (Milnor [1963], p. 14) we know that for ϵ sufficiently small there is a deformation $h_s^2 : M^{c_1+\epsilon} \to M^{c_1-\epsilon} \cup H$, where H is a neighborhood of p_1, small enough so that f can be written in the form of the Morse lemma:

$$f(x) = c_1 + x_1^2 + \cdots + x_{n-\lambda}^2 - y_1^2 - \cdots - y_\lambda^2.$$

Applying this deformation to γ_1 to get $\gamma_2 \subseteq M^{c_1-\epsilon} \cup H$. Note that the deformations h_s^1, h_s^2 are constructed to leave M^a pointwise fixed, so the deformations do not move the endpoints of γ. Let $L \subset H$ be the set on which $y_1, \ldots, y_\lambda$ are zero. It is a standard result of transversality theory (see Sternberg [1964], for example) that given manifolds I and $L \subseteq M$ and a map $\gamma_2 : I \to M$, then if $\dim I + \dim L < \dim M$, γ_2 can be perturbed by an arbitrary small amount so that $\gamma_2(I) \cap L = \emptyset$. Thus in the above (since $\lambda \geq 2$) we can perturb γ_2 so that it does not meet L, and since we need only change γ_2 at those points where its image lies near L, we can still leave the endpoints fixed.

There is a third deformation $h_s^3 : (M^{c_1 - \epsilon} \cup H) - L \to M^{c_1 - \epsilon}$, which one performs by deforming H into points of the form $(0, 0, \ldots, y_1, \ldots, y_\lambda)$ (simply contract the x-coordinates to zero) and then projecting from the center of the subspace of the y-coordinates. Since only points in L have y-coordinates all zero, $H - L$ can be deformed in this fashion and the deformation extends to M^{c_2} as well (see Milnor [1963], pp. 17–19).

Applying this deformation to γ_2, we get $\gamma_3 \subseteq M^{c_2 - \epsilon}$. Continuing inductively we get $\gamma_3 \subseteq M^{c_k - \epsilon}$. But since f has no critical points in $M^{c_k - \epsilon} - M^a$, we can deform $M^{c_k - \epsilon}$ to M^a and thus get the theorem in this case. (See Fig. 4.3).

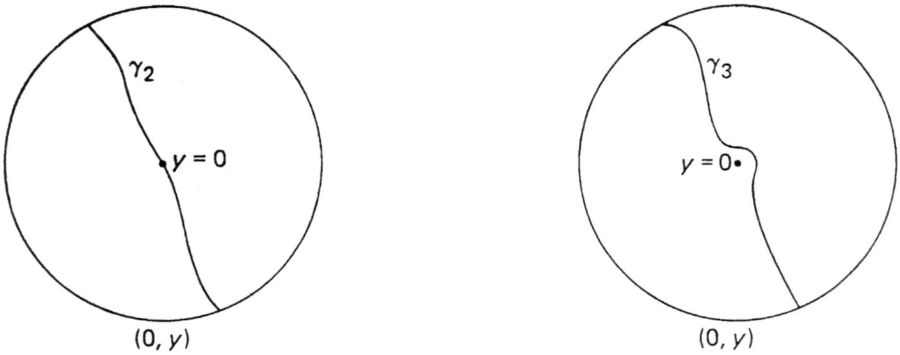

Fig. 4.3.

If f has degenerate critical points, approximate f by $\hat{f}$ as in Milnor [1963], p. 32. Since the index is lower semicontinuous, we may apply the previous argument to $\hat{f}$.

(2) If p_1 and p_2 are nondegenerate critical points of index zero, by Lemma 4.4 there exist distinct coordinate balls B_1 and B_2 about p_1 and p_2 such that f restricted to the boundary of B_i is bounded below by $f(p_i) + 2\epsilon$. But if we let γ be a curve from p_1 to p_2, by (1), γ can be deformed to $\hat{\gamma}$, where $f|\hat{\gamma} \leq \max\{f(p_1), f(p_2)\} + \epsilon$. Since $\hat{\gamma}$ must cross ∂B_1 and ∂B_2, we have a contradiction and the lemma follows. □

Now we turn to the most important use of Morse Theory in Riemannian geometry, namely, its application to the space of curves on a manifold. Again we will merely summarize the results appearing in Milnor [1963].

Let $\Omega(p, q)$ be the set of continuous piecewise smooth curves in M from p to q; that is, $\gamma \in \Omega(p, q)$ if $\gamma : [0, 1] \to M$ is continuous, $\gamma(0) = p$, $\gamma(1) = q$, and there exist $0 = t_0 < t_1 < \cdots < t_n = 1$ such that $\gamma|[t_i, t_{i+1}]$ is smooth. We define a metric on $\Omega(p,q)$ as follows: Let γ, γ_0 be two curves in $\Omega(p,q)$, and let $L(t), L_0(t)$ be functions denoting the arclength of γ (respectively γ_0) from $\gamma(0)$ to $\gamma(t)$. then $d(\gamma,\gamma_0)$ the distance from γ to γ_0 is defined by

$$d(\gamma,\gamma_0) = \max_{0\le t\le 1}\Big\{\rho(\gamma(t),\gamma_0(t)) + \Big[\int_0^1 \Big(\frac{\mathrm{d}L}{\mathrm{d}t} - \frac{\mathrm{d}L_0}{\mathrm{d}t}\Big)^2 \mathrm{d}t\Big]^{\frac{1}{2}}\Big\}.$$

It is easy to see that d is a metric and that the energy function

$$E(\gamma) = \int_0^1 \Big(\frac{\mathrm{d}L}{\mathrm{d}t}\Big)^2 \mathrm{d}t$$

is continuous from $\Omega(p,q)$ to $\mathbb{R}$. Note that if T is the tangent vector to γ,

$$E(\gamma) = \int_0^1 ||T||^2 \mathrm{d}t.$$

Also, by the Schwarz inequality, we have

$$(L[\gamma])^2 = \Big(\int_a^b ||T||\mathrm{d}t\Big)^2 \le \int_a^b ||T||^2\mathrm{d}t \int_a^b 1^2\mathrm{d}t = E(\gamma)(b-a),$$

with equality if and only if $||T||$ is constant.

We would like to investigate the topology of $\Omega(p,q)$ by studying the critical points of E. However, since $\Omega(p,q)$ is not a finite-dimensional manifold, we use an indirect approach.[1]

Let

$$\Omega^c(p,q) = \{\gamma \in \Omega(p,q) | E(\gamma) \le c\}.$$

We shall cunstruct a subspace $\Omega(t_0,\ldots,t_k)^c$ of $\Omega(p,q)^c$ which is a deformation retract of $\Omega^c(p,q)$, and is also a manifold. Then we will examine $E|\Omega(t_0,t_1,\ldots,t_k)^c$.

Given c, let $r_c(M)$ be chosen such that for any point m with $\rho(p,m) \le \sqrt{c}\exp_m$ is a diffeomorphism on the ball $B_{r_c}(M) \subset M_m$. Choose $0 = t_0 < t_1 < \cdots < t_k = 1$ such that

$$c^{-1}(r_c(M))^2 \ge t_{i+1} - t_i.$$

Let $\Omega(t_0,\ldots,t_k)$ be a set of paths γ in $\Omega(p,q)$ such that $\gamma|[t_i,t_{i+1}]$ is a geodesic. Let

$$\Omega^c(t_0,\ldots,t_k) = \Omega(t_0,\ldots,t_k) \cap \Omega^c.$$

The broken geodesic γ in $\Omega^c(t_0,\ldots,t_k)$ is uniquely determined by the points $\{\gamma(t_i)\}$.

The map $\gamma \to (\gamma(t_1),\ldots,\gamma(t_{k-1}))$ thus defines an injection of

$$\Omega^c(t_0,\ldots,t_k) \to M \times \cdots \times M \qquad (k-1 \text{ times}).$$

---

[1]It is possible to extend Morse Theory to infinite-dimensional manifolds in such a way that $\Omega(p,q)$ may be treated more directly, see Palais [1969] and Schwartz [1964].

The image of this map will be a submanifold of $M \times M \times \cdots \times M$ with boundary, whose interior corresponds to

$$\{\gamma \in \Omega(t_0, \ldots, t_k) | E(\gamma) < c\}.$$

Therefore we can endow $\Omega^c(t_0, \ldots, t_k)$ with the structure of a manifold, and the function $E|\Omega^c(t_0, \ldots, t_k)$ is a smooth function.

We now proceed to construct a deformation of $\Omega^c(p, q)$ onto $\Omega^c(t_0, \ldots, t_k)$. Let $\gamma \in \Omega^c(p, q)$. Define σ_s^i to be the unique minimal geodesic from $\gamma(t_i)$ to $\gamma(t_i + s(t_{i+1} - t_i))$. Define $h_s(t)$ by

$$h_s(t) = \begin{cases} \sigma_s^i(t), & t_i \le t \le t_i + s(t_{i+1} - t_i), \\ \gamma(t), & t_i + s(t_{i+1} - t_i) \le t \le t_{i+1}. \end{cases}$$

Let

$$\gamma_s^i = \gamma | [t_i, t_i + s(t_{i+1} - t_i)].$$

Then

$$E(\gamma_s^i) s(t_{i+1} - t_i) \ge (L[\gamma_s^i])^2 \ge (L[\sigma_s^i])^2 = E(\sigma_s^i) s(t_{i+1} - t_i).$$

Since $E(\gamma_s^i) \ge E(\sigma_s^i)$, it follows that for all s, $h_s \in \Omega^c(p, q)$. h_s obviously defines a deformation retraction of $\Omega^c(p, q)$ onto $\Omega^c(t_0, \ldots, t_k)$.

Now it remains to discuss the critical points of E. First take a 1-parameter family of curves $\{c_s\} \subseteq \Omega(p, q)$. Then computing as in (1.5), we find

$$\frac{1}{2} \frac{\mathrm{d}}{\mathrm{d}s}(E(c_s))\Big|_{s=0} = -\int_0^1 \langle W, \nabla_T T \rangle \mathrm{d}t + \sum_i \Delta_{t_i} \langle W, T \rangle,$$

where

$$T = \frac{\partial}{\partial t}(c_s(t)), \qquad W = \frac{\partial}{\partial s}(c_s(t)),$$

and

$$\Delta_{t_i} \langle W, T \rangle = \lim_{t \to t_i^+} \langle W, T \rangle - \lim_{t \to t_i^-} \langle W, T \rangle,$$

where $\{t_i\}$ are the points at which c_0 may fail to be smooth. If c_0 is a broken geodesic $\int \langle W, \nabla_T T \rangle = 0$ and c_0 is smooth, then $\sum \Delta_{t_i} \langle W, T \rangle = 0$. In fact, $\mathrm{d}(E(c_t))/\mathrm{d}t = 0$ for all variations of c_0 exactly when c_0 is a non-broken geodesic, and, conversely, the unbroken geodesics are the critical points of $E|\Omega^c(t_0, \ldots, t_k)$. The argument is as in Chapter 1.

We can compute the second variation of E in a similar fashion. Namely, if γ is a geodesic in $\Omega(p, q)$, if u and v are the parameters of a 2-parameter variation of γ, and if V and W are the vector fields on γ induced by the variation, then computing as in (1.18) we get

$$\frac{\partial^2 E}{\partial u \partial v} = \int_0^1 \langle \nabla_T V, \nabla_T W \rangle + \langle R(V, T)W, T \rangle,$$

which we defined in Chapter 1 to be $I(V, W)$, the index form. If we again consider $E|\Omega^c(t_0, \ldots, t_i)$, then γ is varied through broken geodesics, so V and W above will be broken Jacobi fields. Thus the index of a critical point of $E|\Omega^c(t_0,\ldots,t_i)$ is the index of $I(\ ,\ )$ on broken Jacobi fields. It will follow from Lemma 4.17 and 4.18 that this is the same as the index of I on all vector fields.

Proposition 1.20 says in particular that γ is a *degenerate critical point of* E *if and only if* p *is conjugate to* q *along* γ. However, for any M it is possible to find p, q in M which are not conjugate along any geodesic. In fact, it is a consequence of Sard's Theorem that *given* p*, for almost all* q*,* q *is not conjugate to* p *along any geodesic.* For a proof, see Milnor [1963], p. 98. For such p and q, $E|\Omega^c(t_0,\ldots,t_i)$ satisfies the hypothesis of Theorem 4.10. Thus $\Omega^c(t_0,\ldots,t_i)$ has the homotopy type of a CW-complex with one cell of dimension λ for each critical point γ of index λ. We shall soon prove the Morse Index Theorem, which says that the index of γ as a critical point of E is equal to the number of points on γ conjugate to p, counted according to their multiplicity.

Thus Ω^c has the homotopy type of a CW-complex that has a cell of dimension λ for each geodesic γ from p to q of energy $\leq c$ which has λ points conjugate to p.

Since $\Omega = \bigcup_{c\in\mathbb{R}^+} \Omega^c$, an increasing union, we find the following result.

THEOREM 4.12 (Fundamental Theorem of Morse Theory). *If* p *is not conjugate to* q *along any geodesic, then* $\Omega(p,q)$ *has the homotopy type of a CW-complex with one cell of dimension* λ *for each geodesic* γ *from* p *to* q *which contains* λ *points conjugate to* p*. (See Milnor [1963], p. 95.)*

As an interesting example, we apply this theorem to a compact Lie group G. If we endow G with a bi-invariant metric, we know by Proposition 3.41 that all conjugate points have even order. Thus the index of any geodesic is even. Hence, if $p, q \in G$ are not conjugate, $\Omega_{p,q}(G)$ is homotopy equivalent to a CW-complex with only even-dimensional cells. It follows that the integral homology of $\Omega_{p,q}(G)$ is trivial in odd dimensions and free abelian in even dimensions. The rank of $H_\lambda(\Omega_{p,q})$ equals the number of geodesics of index λ; cf. Milnor [1963].

The following will be useful in Chapter 5.

PROPOSITION 4.13. *If* M *and* M_0 *are manifolds with geodesics* γ, γ_0 *of equal length, and if for all* t *the minimum curvature of* M *at* $\gamma(t)$ *is greater than or equal to the maximum curvature of* M_0 *at* $\gamma_0(t)$*, then the index of* γ *must be greater than or equal to the index of* γ_0*. Hence* γ *as a critical point of* E *has greater index than* γ_0*.*

PROOF. Choose a correspondence between vector fields along γ, γ_0, as in the proof of Rauch I. Then the result is immediate from the formula for the second variation of E. □

Let $\gamma : [0,1] \to M$ be a geodesic from p to q. Then the index form I is a symmetric bilinear form on the set of continuous piecewise smooth vector fields $\chi_0(\gamma)$ on γ which vanish at p and q. The following theorem furnishes a good method for computing the index.

THEOREM 4.14 (Morse Index Theorem). *The subspace of $\chi_0(\gamma)$ on which I is negative definite is finite dimensional, and its dimension is equal to the number of conjugate points to p on $\gamma([0,1])$ counted according to their multiplicity. The null space of I is zero unless q is conjugate to p, and in that case its dimension is the order of the conjugate point q.*

PROOF. The proof of this theorem consists of several steps. If V, W are in $\chi(\gamma)$, then

$$\begin{aligned} I(V,W) &= \int_0^1 \langle \nabla_T V, \nabla_T W\rangle - \langle R(T,V)T, W\rangle \\ &= \sum_i \Delta_{t_i}\langle \nabla_T V, W\rangle - \int_0^1 \langle \nabla_T^2 V, W\rangle + \langle R(T,V)T, W\rangle. \end{aligned}$$

In particular, if for all i, $V|[t_i, t_{i+1}]$ is a Jacobi field, then

$$I(V,W) = \sum \Delta_{\lambda_i}\langle \nabla_T V, W\rangle.$$

Now pick an increasing sequence

$$0 = t_0 < t_1 < \cdots < t_n = t \leq 1$$

such that each segment $\gamma|[t_i, t_{i+1}]$ has no pair of conjugate points. Let

$$\chi^1 = \{V \in \chi_0(\gamma) | V(t_i) = 0\}$$

and let $\chi(t_0, \ldots, t_k)$ be the subspace of those vector fields in $\chi_0(\gamma)$ which are Jacobi fields on each segment $[t_i, t_{i+1}]$. Since a Jacobi field on $\gamma|[t_i, t_{i+1}]$ is determined by its values at the endpoints,

$$\chi(t_0, \ldots, t_k) = M_{\gamma(t_1)} \oplus \cdots \oplus M_{\gamma(t_{k-1})}.$$

This identification is exactly the tangent map at γ to the injection

$$\Omega^c(t_0, \ldots, t_k) \to M \times M \times \cdots \times M.$$

Also,

$$\chi_0(\gamma) = \chi^1 \oplus \chi(t_0, \ldots, t_k),$$

where for $V \in \chi_0(\gamma)$, the element of $\chi(t_0, \ldots, t_k)$ is determined by $V(t_1)$, $V(t_2)$, $\ldots$, $V(t_{k-1})$. □

We pause for a moment to prove several preparatory lemmas.

LEMMA 4.15. *$V \in \chi_0(\gamma)$ is in the null space of I if and only if V is a Jacobi field.*

PROOF. This is just a restatement of Proposition 1.20. □

COROLLARY 4.16. *The dimension of the null space of I equals the order of the conjugate point q.*

LEMMA 4.17. *$I|\chi^1$ is positive definite.*

PROOF. Let I_i be the index form on $\gamma|[t_i, t_{i+1}]$. Then $I|\chi^1 = \sum I_i$. Let $W \in \chi^1$. Then by the index lemma,

$$I_i(W, W) \geq I_i(J, J),$$

where J is a Jacobi field such that $J_{t_i} = J_{t_{i+1}} = 0$. But then $J = 0$ on $[t_i, t_{i+1}]$, so $I_i(W, W) \geq 0$. Also, $I_i(W, W) = 0$ only when $W = J = 0$. Therefore,

$$I(W, W) = \sum_i I_i(W, W) \geq 0$$

with equality only when $W = 0$ on all of γ. □

LEMMA 4.18. *χ^1 and $\chi(t_0, \ldots, t_k)$ are perpendicular with respect to I.*

PROOF. Let $W \in \chi(t_0, \ldots, t_k)$, $W \in \chi^1$. The lemma is immediate from the fact that W vanishes at the jumps of $\nabla_T V$, and the formula for I when V is a broken Jacobi field. □

COROLLARY 4.19. *Index(I) = Index($I|\chi(t_0, \ldots, t_k)$). Hence the index is finite.*

PROOF. Use Lemmas 4.17 and 4.18. □

PROOF OF THE MORSE INDEX THEOREM 4.14. By use of the identification of $\chi(t_0, \ldots, t_{k-1}, t)$, for variable t, with

$$M_{\gamma(t_1)} \oplus \cdots \oplus M_{\gamma(t_{k-1})} = \mathfrak{v}.$$

we can view $I|\chi(t_0, \ldots, t_{k-1}, t)$ as a 1-parameter family of bilinear forms I_t on a fixed vector space $\mathfrak{v}$. Let $v \in M_{\gamma(t_{k-1})}$ and $J_{t,v}$ be the Jacobi field on $[t_{k-1}, t]$ such that $J_{t,v}(t_{k-1}) = v$ and $J_{t,v}(t) = 0$. Clearly $J_{t,v}$ varies continuously with t. Hence so does I_t. Moreoever, for $t' > t$, let $\tilde{J}_{t,v}$ be the field on $[t_{k-1}, t']$ which agrees with $J_{t,v}$ on $[t_{k-1}, t]$ and is zero on $[t, t']$. Then by the First Index Lemma,

$$I(\tilde{J}_{t,v}, \tilde{J}_{t,v}) > I(J_{t',v}, J_{t',v}).$$

Thus for fixed $v \in \mathfrak{v}$, $t' > t$ implies $I_{t'}(V, V) < I_t(V, V)$. It follows that the index of $I_{t'}$ for $t' > t$ is at least equal to the index of I_t, plus the nullity of I_t. Because I_t depends continuously on t, for $t' - t$ sufficiently small, the opposite inequality is also true. For t sufficiently small, the index of I_t is zero by the basic index lemma. The theorem now follows from an obvious counting argument and Lemma 4.15. □

REMARK 4.20. There is also an index theorem for focal points. (See Postnikov [1965].)

CHAPTER 5

Closed Geodesics and the Cut Locus

In Corollary 1.9, we observe that for each p the exponential map is injective on a sufficiently small ball in M_p. Hence we take up the question of determining the maximum radius of such a ball. This is definitely a global problem. The results of the chapter will be applied in Chapters 6 and 7.

Let $\gamma : [0, \infty) \to M$ be a geodesic with $\gamma(0) = m$. All geodesics in this chapter will be parameterized by arc length, unless we make explicit mention to the contrary, and all manifolds are assumed to be *complete*. For sufficiently small $t > 0$ we know that $\rho(\gamma(t), \gamma(0)) = t$, since $\exp_m$ is injective on a sufficiently small ball in M_m. On the other hand if $\rho(\gamma(t_1), \gamma(0)) < t_1$, then for any $\alpha > 0$

$$\rho(\gamma(t_1 + \alpha), \gamma(0)) \leq \rho(\gamma(t_1 + \alpha), \gamma(t_1)) + \rho(\gamma(t_1), \gamma(0)) < t_1 + \alpha.$$

It follows that the set of t for which $\rho(\gamma(t), \gamma(0)) = t$ is either $[0, \infty)$ or $[0, t_0)$ for some $t_0 > 0$. In the latter case, $\gamma(t_0)$ is called the *cut point* of γ with respect to m. The union of all cut points is called the *cut locus* of m and denoted by $C(m)$. If M is compact, it follows that every γ must have a cut point. Conversely if for some $m \in M$, every γ has a cut point, then M is necessarily compact.

EXAMPLE 5.1. $M = S^n$. In this case, for any $m \in S^n$, $C(m)$ is the antipodal point $\overline{m}$.

The following gives a useful characterization of cut points.

LEMMA 5.2. *$\gamma(t_0)$ is a cut point of $m = \gamma(0)$ along γ if and only if one of the following holds for $t = t_0$ and neither holds for any smaller value of t:*

(a) $\gamma(t_0)$ is conjugate to m along γ, or
(b) there is a geodesic $\sigma \neq \gamma$ from m to $\gamma(t_0)$ such that $L[\sigma] = L[\gamma]$.

PROOF. Let $\epsilon_i > 0$ be a sequence with $\epsilon_i \to 0$. Let $\sigma_i \neq \gamma$ be a sequence of minimal geodesics from m to $\gamma(t_0+\epsilon_i)$. By compactness of the unit sphere in $T_m(M)$ we may assume that there is a geodesic σ such that $\sigma_i'(0) \to \sigma'(0)$ for some subsequence. By continuity, $L[\sigma] = L[\gamma]$, and either $\sigma \neq \gamma$ or, by the inverse-function theorem, $d\exp_m$ must be singular at $t_0\gamma'(0)$.

As we know, geodesics do not minimize past the first conjugate point. Similarly, if there exists $\sigma \neq \gamma$ from $\gamma(0)$ to $\gamma(t_0)$, then the curve $\sigma|[0, t-\epsilon] \cup \tau$ has length $< t + \epsilon$, if τ is the nimimal geodesic from $\sigma(t - \epsilon)$ to $\gamma(t + \epsilon)$. □

COROLLARY 5.3. *If $\gamma(t_0)$ is a cut point of $\gamma(0)$ along γ, then $\gamma(0)$ is a cut point of $\gamma(t_0)$ along $-\gamma$.*

PROPOSITION 5.4. *The distance to the cut point is a continuous function defined on an open subset of the unit sphere bundle of M. In particular, C(m) is a closed set.*

PROOF. Let $\gamma_i(0) \to \gamma(0)$, $\gamma_i'(0) \to \gamma'(0)$, and let t_0^i, t_0 be distances to the cut points of γ_i, γ (with possibly $t_0^i = \infty$). If for infinitely many γ_i and fixed t, $\rho(\gamma_i(t), \gamma_i(0)) = t$, it follows by continuity of the distance function that $\rho(\gamma(t), \gamma(0)) = t$ and hence that the dependence is upper semicontinuous. Let $\bar{t} = \limsup t_0^i$. Since the set of singular values of the exponential map is closed, an accumulation point of conjugate points is a conjugate point. We may therefore assume that there is a sequence $t_0^i \to \bar{t}$ such that there exist $\sigma_i \neq \gamma_i$, with

$$\sigma_i(0) = \gamma_i(0), \qquad \sigma_i(t_0^i) = \gamma_i(t_0^i), \qquad L[\sigma_i] = L[\gamma_i], \qquad \sigma_i \to \sigma.$$

We may also assume that $\gamma(\bar{t})$ is not conjugate to $\gamma(0)$. Hence by the inverse function theorem there is a neighborhood U of $\bar{t}\gamma'(0)$ in $T(M)$ on which $\exp_p|U$ is locally one to one for all $p \in \Pi(U)$, where $\Pi : T(M) \to M$. It follows that $\sigma \neq \gamma$, and hence that the above dependence is lower semicontinuous. □

It follows from Proposition 5.4 that for each p there exists $q \in C(p)$ such that $\rho(p, q) = \rho(p, C(p))$ provided $C(p) \neq \emptyset$.

Let

$$\mathfrak{G} = \{v \in M_m | \rho(\exp_m v, m) = ||v||\}.$$

The boundary $\partial\mathfrak{G}$ is called the *cut locus in the tangent space* M_m. By what we have seen above, if M is compact, $\mathfrak{G}$ is homeomorphic to a closed ball and M is obtained from this ball by making certain identifications on the boundary. Note also that $M - m$ is homotopy equivalent by radial projection to the cut locus $C(m)$. Now $\exp_m|\mathfrak{G} - \partial\mathfrak{G}$ is 1-1 onto $M - C(m)$, and hence $M - C(m)$ is an open ball. It is important to be able to estimate a lower bound for the size of this ball (or, more precisely, $\min_{v \in \partial\mathfrak{G}}(||v||)$ in terms of geometric properties of M. Roughly speaking, this allows one to distinguish between local and global situations in M.

DEFINITION 5.5. The *injectivity radius i(M)* of M is the largest r such that for all m, $\exp_m$ is an embedding on the open ball of radius r in M_m. Alternatively

$$i(M) = \min\{||v|| \, |m \in M, v \in \partial\mathfrak{G}(m)\}.$$

As above, it follows from Proposition 5.4 that if M is compact, there exist $p, q \in C(p)$ such that $\rho(p, q) = i(M)$.

LEMMA 5.6. *Let q realize the minimum distance from p to its cut locus. Then either there is a minimal geodesic γ from p to q along which q is conjugate to p, or there are precisely two minimal geodesics γ, σ from p to q such that $\gamma'(p) = -\sigma'(p)$.*

PROOF. By Lemma 5.2, we may assume that there exist distinct minimal geodesics γ, σ such that q is not conjugate to p along either. If $\gamma'(q) \neq -\sigma'(q)$, then there must exist $v \in M_q$ such that

$$\langle v, -\gamma'(q)\rangle > 0, \qquad \langle v, -\sigma'(q)\rangle > 0.$$

Let τ be a geodesic such that $\tau'(0) = v$. Since q is not conjugate to p along γ or σ, for sufficiently small s there exist 1-parameter families of geodesic segments γ_s, σ_s going from p to $\tau(s)$ with $\gamma_0 = \gamma, \sigma_0 = \sigma$.

By the first variation formula,

$$\frac{\mathrm{d}}{\mathrm{d}s}(L[\gamma_s])\Big|_{s=0} < 0, \qquad \frac{\mathrm{d}}{\mathrm{d}s}(L[\sigma_s])\Big|_{s=0} < 0,$$

and hence for sufficiently small s, we have say

$$L[\gamma_s] \leq L[\sigma_s] < L[\gamma] = L[\sigma].$$

It follows from Lemma 5.2 that there is a cut point of σ_s between p and $\tau(s)$. Since $L[\sigma_s] < L[\gamma]$, this is a contradiction. □

COROLLARY 5.7 (Klingenberg). *If $K \geq K_M \geq H > 0$, then*

$$i(M) \geq \min\Big\{\frac{\pi}{\sqrt{K}}, \textit{half the length of the shortest smooth closed geodesic in } M\Big\}.$$

PROOF. We can assume that there exist $p, q \in C(p)$ such that $\rho(p,q) = i(M)$ and q is not conjugate to p along any minimal geodesic. Then there exist precisely two geodesics γ, σ such that

$$\gamma(0) = \sigma(0) = p, \qquad \gamma(t_0) = \sigma(t_0) = q, \qquad \gamma'(t_0) = -\sigma'(t_0).$$

Now $q \in C(p)$ implies $p \in C(q)$, and clearly p, which is not conjugate to q, is a point of $C(q)$ closest to q. Hence $\sigma'(0) = -\gamma'(0)$, and γ and σ fit together to form a smooth closed geodesic. If, on the other hand, q is conjugate to p, then by Corollary 1.35, $\rho(p,q) > \frac{\pi}{\sqrt{K}}$. □

We now derive some results on finding a lower bound for $i(M)$ in various situations. First, we will give a completely general theorem which is not as sharp as possible in the special situations handled by the theorems which follow it. However, if one of the bounds is removed and no other restrictions are added then counterexamples exist.

Let $d(M)$ denote the diameter of M and $V(M)$ the volume; see Kobayashi and Nomizu [1963, 1969].

THEOREM 5.8 (Cheeger [1970]). *Given $n, d, V > 0$ and H, there exist $c_n(d, V, H) > 0$ such that if M is an n-dimensional Riemannian manifold such that $d(M) \leq d$, $V(M) > V$ and $K_M \geq H$, then every smooth closed geodesic on M has length $> c_n(d, V, H)$.*

PROOF. The idea is to show that the existence of a sufficiently short closed geodesic would imply $V(M) \le V$. As a matter of notation, in a vector space with inner product let $\mathfrak{a}_{t,\theta}(v)$ (respectively $\mathfrak{a}'_{t,\theta}(v)$) denote the set of vectors of length $\le t$ making an angle $\alpha \le \theta < \frac{1}{2}\pi$ (respectively $\theta \le \alpha \le \frac{1}{2}\pi$) with either v or $-v$. Let M^H denote the canonical simply connected space of constant curvature H and dimension n. Define θ^* by the condition

$$V(\exp_m(\mathfrak{a}'_{d,\theta^*}(v))) = \frac{1}{2}V,$$

where $m \in M^H$ and $v \in M^H_m$. Clearly $\theta^* < \frac{1}{2}\pi$. Now define r by

$$V(\exp_m(\mathfrak{a}_{r,\theta^*}(v))) = \frac{1}{2}V.$$

(If $H > 0$, it follows easily from Rauch I that $r < \frac{1}{2}\pi/\sqrt{H}$, since in that case $\frac{1}{2}V < \frac{1}{2}V(M) \le \frac{1}{2}V(M^H)$.) Now let σ, τ be geodesics in M^H such that

$$\sigma(0) = \tau(0), \qquad \sphericalangle(\sigma'(0), \tau'(0)) = \alpha \le \theta^* < \frac{1}{2}\pi.$$

Since $\sigma|[0,r]$ is minimal, it follows from the first variation formula that $\rho(\sigma(r), \tau(t))$ is strictly decreasing for sufficiently small t. Using the continuity of the distance function, and the fact that θ^* is strictly less than $\frac{1}{2}\pi$, we deduce the existence of a number $c_n(d, H, V) > 0$ such that for any σ, τ as above, with $\alpha \le \theta^*$, $\rho(\sigma(r), \tau(t)) < r$ provided $t \le c_n(d, H, V)$. We write $c_n(d, H, V)$ because this number clearly depends only on n, H and r, where r depends only on n, H, d, V.

We now show that $c_n(d, H, V)$ has the property claimed in the theorem. Let γ be a smooth closed geodesic in M of length $l \le c_n(d, H, V)$. It will suffice to show that the cut locus in $M_{\gamma(0)}$ is contained in $\mathfrak{a}'_{d,\theta^*}(\gamma'(0)) \cup \mathfrak{a}_{r,\theta^*}(\gamma'(0))$. This region is depicted in Fig 5.1. Then by Rauch I, we have

$$\begin{aligned} V < V(M) &\le V(\exp_{\gamma(0)}(\mathfrak{a}'_{d,\theta^*}(\gamma'(0)))) + V(\exp_{\gamma(0)}(\mathfrak{a}_{r,\theta^*}(\gamma'(0)))) \\ &\le V(\exp_m(\mathfrak{a}'_{d,\theta^*}(v))) + V(\exp_m(\mathfrak{a}_{r,\theta^*}(v))) = \frac{1}{2}V + \frac{1}{2}V = V. \end{aligned}$$

which is a contradiction.

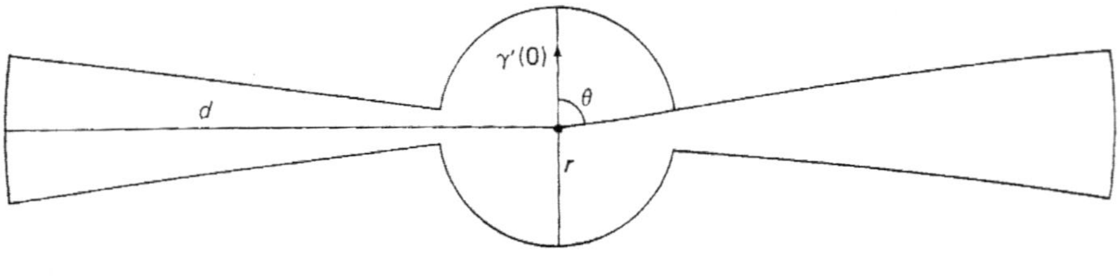

Fig. 5.1.

Since $d(M) \leq d$, it suffices to show that if $w \in \mathfrak{a}_{r,\theta^*}(\gamma'(0))$, then $\exp_{\gamma(0)}tw$ does not minimize beyond $t = r$. Now either $\sphericalangle(w, \gamma'(0)) \leq \theta^*$ or $\sphericalangle(w, -\gamma'(0)) \leq \theta^*$. Suppose the former. Toponogov's Theorem 2.2 and the construction of $c_n(d, H, V)$ imply that $\rho(\exp_{\gamma(0)}rw, \gamma(l)) < r$. (We can apply Toponogov because $c_n(d, H, V) < \pi/\sqrt{H}$) and if $\exp_{\gamma(0)}tw|[0, r]$ is not minimal, there is nothing to prove. (The fact that $\dim M^H$ may be > 2 is of no importance, since σ, τ above span a 2-dimensional totally geodesic submanifold.) Since γ is closed, $\gamma(l) = \gamma(0)$. Thus $\rho(\exp_{\gamma(0)}rw, \gamma(0)) < r$, which implies that $\exp_{\gamma(0)}tw$ does not minimize past $t = r$. Since γ is smooth and closed,

$$-\gamma'(0) = -\gamma'(l).$$

Therefore, if

$$\sphericalangle(w, -\gamma'(0)) = \sphericalangle(w, -\gamma'(l)) \leq \theta^*,$$

the same argument applies. □

Note that in case the Ricci curvature of M satisfies

$$\mathrm{Ric}(x, x) \geq (n-1)H > 0$$

for all unit vectors x, then by Meyer's Theorem, $d(M) \leq \pi\sqrt{H}$. Also, if the Euler characteristic or some Pontryagin number $\#(M)$ is nonzero and B denotes the corresponding characteristic form (see Chern [1944], Kobayashi and Nomizu [1963, 1969]),

$$\int B = \#(M), \qquad \int ||B|| \mathrm{d}A \geq |\#(M)| \geq 1,$$

then

$$V(M) \geq (\max||B||)^{-1}.$$

Now $\max||B||$ may be estimated from above in terms of $\max||K_M||$. Thus if these conditions are satisfied, Theorem 5.8 is particularly useful.

THEOREM 5.9. *Let M be even-dimensional orientable and $K \geq K_M > 0$. Then:*

(1) (Klingenberg) *If p,q are such that $q \in C(p)$ and $\rho(p, q) = i(M)$, then q is conjugate to p. Hence $i(M) \geq \pi/\sqrt{K}$.*

(2) (Synge) *M is simply connected.*

PROOF. (1) The proof involves a quite typical application of the second variation formula. By Corollary 5.7, we may assume that there is a smooth closed normal geodesic γ with $\gamma(0) = p$ through p and q such that $L[\gamma] = l = 2i(M)$. By parallel translating around γ we obtain an isometry $P_\gamma : M_{\gamma(0)} \to M_{\gamma(0}$ which is orientation preserving and such that $P_\gamma(\gamma'(0)) = \gamma'(0)$. Hence the restriction of P_γ to the subspace $N \subset M_{\gamma(0)}$ normal to $\gamma'(0)$ is an orientation-preserving isometry of N to itself. Since M is even dimensional, N is odd dimensional, and it follows from elementary linear algebra that $P_\gamma|N$ must leave some vector $v \in N$ invariant. Hence v extends to a parallel vector field $V(t)$ normal to γ such that $V(l) = V(0)$. Choose a smooth variation of γ by closed curves h_s such that $h_0 = \gamma$ and such that the variation vector field is V (e.g. $h_s(t) = \exp_{\gamma(t)} sV(t)$). Apply the second variation formula. Using the fact that $V(0) = V(l)$, we see that the boundary terms cancel, and we get

$$l''(0) = \int_0^l (-K(V, \gamma')\mathrm{d}t < 0$$

since $V' = 0$ and $K > 0$. Therefore $L[h_s]$ 1has a strict local maximum at $s = 0$.

Let q_s be a point on h_s farthest from $h_s(0)$, and let σ_s be a minimal geodesic from q_s to $h_s(0)$. Since $\rho(h_s(0), q_s) < i(M)$, σ_s is in fact unique and q_s is not conjugate to $h_s(0)$ along σ_s. It follows from the first variation formula that $\sigma_s'(0) \perp h_s'(0)$. Now, since q is the unique point on γ farthest from p, it follows by continuity that $\lim_{s\to 0} q_s = q$. By (local) compactness of the unit sphere bundle, there exists an accumulation point $w \in M_q$ of $\sigma_s'(0)$. By continuity of the distance function and the exponential map, if σ is the geodesic with $\sigma'(0) = w$, then $\sigma|[0, \frac{1}{2}l]$ is a minimal segment from q to p. Clearly $\sigma_s'(0) \perp \gamma'$, and thus σ does not coincide with either segment of γ. But since p realizes the distance from q to $C(q)$, by Lemma 5.6, q is conjugate to p.

(2) We show that if M is compact, each nontrivial free homotopy class $\mathfrak{h}$ of loops contains a member of minimal length which is a smooth closed geodesic. In fact, let l be the infimum of the lengths of curves in some nontrivial class, and let $0 = t_0 < \cdots < t_n = 1$ be a subdivision of $[0, 1]$ such that $t_{i+1} - t_i < i(M)/l$ for all i. Let γ_j be a sequence of curves parameterized proportional to arc length such that

$$\lim_{i\to\infty} L[\gamma_i] = l = \inf_{h\in\mathfrak{h}} L[h].$$

By the argument of Chapter 4, we may assume that each γ_j is a broken geodesic with breaks at $\gamma_j(t_i)$. By compactness of M, we may assume that the sequence $\gamma_j(t_i)$ converge. Hence $\{\gamma_j\}$ approaches some limit curve γ which is clearly in the same free homotopy class and is a geodesic with at most breaks at $\gamma(t_i)$. But since γ has minimal length, an easy application of the first variation formula shows that γ is actually smooth. However, the argument of (1) shows that a smooth closed geodesic can be deformed to shorter curves, which contradicts the minimality of γ. Hence there exists no nontrivial free homotopy class and M is simply connected. □

In case M is odd dimensional, simply connected, and has curvature satisfying $K \geq K_M \geq \frac{1}{4}K > 0$, then the first part of the above result still holds. We shall only treat the case $K_M > \frac{1}{4}K$. The case $K_M \geq \frac{1}{4}K$ is similar but technically more complicated (see Cheeger and Gromoll [1972]). The proof does not use explicitly the fact that M is odd dimensional, only that $\dim M \geq 3$.

THEOREM 5.10 (Klingenberg). *Let M be a simply connected Riemannian manifold of* $\dim \geq 3$ *such that* $K \geq K_M \geq H > \frac{1}{4}K > 0$. *Then* $i(M) \geq \pi/\sqrt{K}$.

PROOF. By Corollary 5.7, it suffices to show that any smooth closed geodesic has length $\geq 2\pi/\sqrt{K}$. In fact, let γ be any closed geodesic of length $< 2\pi/\sqrt{K}$, not necessarily smooth. Since M is simply connected, there is a homotopy H_s such that $H_1 = \gamma$ and $H_0 \equiv \gamma(0)$. As in Chapter 4, we may assume that the H_s are piecewise smooth and parameterized proportional to arc length, i.e. we may view H_s as a curve in some finite-dimensional approximation to $\Omega(\gamma(0), \gamma(0))$. Now by Proposition 4.13, comparing with a sphere of curvature H, every geodesic in $\Omega(p, q)$ of length $\geq \pi/\sqrt{H}$ has index at least $n - 1 \geq 2$. Hence by Lemma 4.11 we may replace H_s by a homotopy (also denoted by H_s) such that for any fixed s,

$$L[H_s] \leq \pi/\sqrt{H} + \epsilon < 2\pi/\sqrt{K} - 2\epsilon_1$$

for sufficiently small ϵ, ϵ_1. We now show how this implies that γ can be lifted to a *closed* curve in $M_{\gamma(0)}$. Since the lift of a geodesic is always a ray, this will give a contradiction. Let s_0 be such that $s_i \to s_0$ and the curves H_{s_i} can each be lifted to closed curves

$$\tilde{H}_{s_i} \subset B_{\pi/\sqrt{K}-\epsilon_1}(0) \subset M_{\gamma(0)}$$

(open ball). Then by Ascoli's Theorem, it follows that there is a convergent subsequence $\tilde{H}_{s_{i_j}} \to \tilde{H}_{s_0}$, where $\tilde{H}_{s_0}$ is a closed curve lifting H_{s_0}. We claim that $\tilde{H}_{s_0}$ is also contained in the *open* ball $B_{\pi/\sqrt{K}-\epsilon_1}(0)$. If not, by the Gauss Lemma we could find $t_0 \leq t_1$ such that

$$\int_0^{t_0} ||\tilde{H}'_{s_0}(t)_r|| dt + \int_{t_1}^1 ||\tilde{H}'_{s_0}(t)_r|| dt \geq 2\pi/\sqrt{K} - 2\epsilon_1,$$

where $(\ )_r$ denotes the radial component. But $||\tilde{H}'_{s_0}(t)_r|| \leq ||\tilde{H}'_{s_0}(t)||$, so the above inequality would imply

$$L[H_{s_0}] \geq 2\pi/\sqrt{K} - 2\epsilon_1 \geq \pi/\sqrt{H} - \epsilon_1,$$

which is a contradiction. Thus the set of values for which H_s may be lifted to $B_{\pi/\sqrt{K}-\epsilon_1}(0)$ is closed. Let s_0 be such that H_{s_0} may be lifted to $B_{\pi/\sqrt{K}-\epsilon_1}(0)$. Since exp is locally a diffeomorphism on this ball (see the remarks following Theorem 1.34), by compactness there exists r, s such that, for any s such that $|s_0 - s| < s$, each point $H_s(t)$ has a *unique* inverse image $\tilde{H}_s(t)$ in the ball $B_r(\tilde{H}_{s_0}(t))$. For sufficiently small r, it is straightforward to check that the $\tilde{H}_s(t)$ define a continuous closed lift of $H_s(t)$ which is contained in $B_{\pi/\sqrt{K}-\epsilon_1}(0)$. Thus the set of s such that H_s may be lifted is open and closed, so by the connectedness of [0,1], γ itself can be lifted, which is our final contradiction. □

One might conjecture that by use of a different argument the hypothesis $K_M \geq \frac{1}{4}K$ could be removed as in the even-dimensional case. The Berger Example 3.35 shows that this is not possible, and that the best one could hope to prove is $L[\gamma] \geq 2\pi/\sqrt{K}$ if $K \geq K_M \geq \frac{1}{9}K$.

Another application of Lemma 4.11 is contained in the following more special result.

THEOREM 5.11. *If M is simply connected and if, for $m \in M$, the first conjugate point along every geodesic γ with $\gamma(0) = m$ is of order ≥ 2, then all geodesics from m minimize up to the first conjugate point.*

PROOF. Let $\gamma|[0,t]$ have index zero and $\gamma(t)$ be not conjugate to $\gamma(0)$ along γ. Since the set of singular points of $\exp_{\gamma(0)}$ is closed, it follows from the Morse Index Theorem that there exist ϵ_1, ϵ_2 such that if $\sphericalangle(\sigma'(0), \sigma'(0)) < \epsilon_1$ and $\epsilon < \epsilon_2$, then σ is of index zero on $[0, t+\epsilon]$. Now by Sard's Theorem (Sternberg [1964]), there exists a sequence of points $p_i \to \gamma(t)$ such that every geodesic from m to p_i has nonconjugate endpoints and hence is of index 0 or ≥ 2. By Lemma 4.11, then for each p_i there is just one geodesic segment γ_i of index zero from m to p_i, and hence γ_i must be the unique minimal segment from m to p_i. However, since $d\exp_m$ is nonsingular at $t\gamma'(0)$, it maps some small neighborhood of $t\gamma'(0)$ onto a neighborhood of $\gamma(t)$. Thus for $\rho(p_i, \gamma(t))$ sufficiently small, there exist segments $\bar{\gamma}_i \to \gamma$ from m to p_i which by the above are of index zero. By uniqueness of the γ_i we have $\bar{\gamma}_i = \gamma_i$ and hence $\gamma_i \to \gamma$. Since γ is the limit of segments which are minimal on $[0,t]$, it must itself be minimal on $[0,t]$. □

COROLLARY 5.12. *If G is a simply connected Lie group with bi-invariant metric, then all geodesics minimize up to the first conjugate point.*

PROOF. This is immediate from Proposition 3.41 and Theorem 5.11. □

More generally:

THEOREM 5.13. *Let $M = G/H$ be a symmetric space with metric induced from the Killing form of G, where G is a compact simply connected semi-simple Lie group and H is the fixed point set of an involutive automorphism σ. Then:*

(1) geodesics minimize up to the first conjugate point in M;

(2) M is simply connected;

(3) H is connected.

PROOF. (1) By Corollary 5.12, we know that geodesics in G minimize up to the first conjugate point, By Proposition 3.42, the map $\Phi : G/H \to G$, $\Phi([gH]) = g\sigma(g^{-1})$ embeds M as a totally geodesic submanifold of G in such a way that the induced metric is 2 times the normal metric on M. It will suffice then to check that if $\gamma : [0, \infty) \to M$ is a geodesic, then the first conjugate point occurs at the same value of t as it does for $\Phi(\gamma) \subset G$. We may assume $\Phi(\gamma(0)) = e$. Recall from Chapter 3 that we have the decompositions

$$\mathfrak{g} = \mathfrak{p} \oplus \mathfrak{h}, \qquad [\mathfrak{h}, \mathfrak{h}] \subset \mathfrak{p}, \qquad [\mathfrak{h}, \mathfrak{p}] \subset \mathfrak{p}, \qquad [\mathfrak{p}, \mathfrak{p}] \subset \mathfrak{h}$$

where $\mathfrak{h}$ is the Lie algebra of H and $\mathfrak{p}$ is identified with $\Phi(M)_e$. Now by the remark preceeding Proposition 3.4 we know that the conjugate points of γ in G occur at values π/λ_i, where the corresponding Jacobi field is of the form $\sin(\sqrt{(\lambda_i t)})E_i$, and E_i is the parallel translate of an eigenvector of $-\frac{1}{4}\mathrm{Ad}^2(T)$, where $T = \mathrm{d}\Phi(\gamma'(0))$. Let v be an eigenvector with eigenvalue λ_i. Since $T \in \mathfrak{p}$, and above relations show that

$$-\frac{1}{4}\mathrm{Ad}^2(T)(\mathfrak{h}) \subset \mathfrak{h}, \qquad -\frac{1}{4}\mathrm{Ad}^2(T)(\mathfrak{p}) \subset \mathfrak{p}.$$

Hence $v = v_{\mathfrak{p}} + v_{\mathfrak{h}}$, $v_{\mathfrak{p}} \in \mathfrak{p}$, $v_{\mathfrak{h}} \in \mathfrak{h}$ and $v_{\mathfrak{p}}, v_{\mathfrak{h}}$ are also eigenvectors. It follows that either $\mathfrak{p}$ or $\mathfrak{h}$ contains a nonzero eigenvector. In fact, $\mathrm{Ad}(T)$ is easily seen to provide an isomorphism between the set of eigenvectors in $\mathfrak{p}$ and those in $\mathfrak{h}$. We conclude that the first conjugate point occurs in M at the same value of t as in G, and in fact with one half of the order.

(2) Assume that M is not simply connected, and let $\gamma : [0, t_0] \to M$ be a minimal closed geodesic in some nontrivial homotopy class. Then there exists $t_1 \leq \frac{1}{2}t_0$ such that $\gamma(t_1)$ is the cut point of γ. So by (1), $\gamma(t_1)$ must be a conjugate point. But in this case γ can be deformed to a shorter curve (as in proof of Corollary 1.29), which is impossible.

(3) $G \to G/H = M$ is a fibration, so we have the exact homotopy sequence (Spanier [1966])

$$\cdots \to \Pi_1(M) \to \Pi_0(H) \to \Pi_0(G) \to \dots$$

Since G is connected and M is simply connected, it follows that H is connected. □

The above result is originally due to Crittenden [1962], where it is proved differently. Our proof is taken from Cheeger [1969].

Appendix

As set $A \subset M$ is called *strongly convex* if its closure $\bar{A}$ has the property that for any $q, q' \in \bar{A}$ there is a unique minimal geodesic $\tau_{q,q'}$ from q to q' such that the interior of $\tau_{q,q'}$ is contained in A. We are going to show that a sufficiently small open metric ball in any Riemannian manifold is strongly convex.

We have seen the distance of a point q to its cut locus is a continuous function of q. Let $B_r(p)$ be contained in some compact set C. Let $\mathfrak{J}$ denote the infimum of this function over C. Let K denote the supremum of sectional curvatures at points of C. In what follows interpret $\pi/\sqrt{K}$ as infinity if $K \leq 0$.

THEOREM 5.14 (cf. Whitehead [1932]). *If $r < \frac{1}{2}\min\{\pi/\sqrt{K}, \mathfrak{J}\}$, then $B_r(p)$ is strongly convex. In particular, there exists a positive continuous* *function* $r(p)$*, the convexity radius, such that* $r < r(p)$ *implies* $B_r(p)$ *is strongly convex.*

The proof of Theorem 5.14 depends on the following, more local, statement.

LEMMA 5.15. *Let* $B_r(p)$ *be a normal coordinate ball such that* $r < \frac{1}{2}\pi/\sqrt{K}$*. Let* $\tau : [0,1] \to B_r(p)$ *be a geodesic segment. Then the function* $\rho(\tau(t), p)$ *has at most one critical point for* $0 < t < 1$*, and such a critical point must be a minimum.*

PROOF. The uniqueness follows easily from the statement that any critical point t_0 is a minimum. Let $\sigma_q : [0,1] \to M$ denote the unique minimal geodesic from p to q. Set $L[\sigma_{\tau(t_0)}] = l$ and consider the variation $\sigma_{\tau(t)}(s)$ of $\sigma_{\tau(t_0)}$. By the second variation formula (1.17)

$$\frac{\mathrm{d}^2}{\mathrm{d}t^2} L[\sigma_{\tau(t)}] = \int_0^l (\langle \nabla_T V, \nabla_T V\rangle - \langle R(T,V)V, T\rangle)\mathrm{d}t.$$

Here the endpoint terms $\langle \nabla_V V, T\rangle$ vanish because τ is a geodesic and the terms $(T\langle V,T\rangle)^2$ vanish because the variation is through geodesics and $\langle V, T\rangle$ vanish at the endpoints. If $K \leq 0$, the integrand is clearly positive. In the space of constant curvature $K > 0$, let $\bar{V}$ be a vector field along some geodesic $\bar{\sigma}$ with the same expression in parallel frames as V. Then by the Index Lemma 1.24,

$$\begin{aligned}
\int_0^l (\langle \nabla_T V, \nabla_T V\rangle - \langle R(T,V)V, T\rangle) &= \int_0^l (\langle \nabla_T V, \nabla_T V\rangle - K(T,V)\langle V, V\rangle)\\
&\geq \int_0^l (\langle \nabla_T V, \nabla_T V\rangle - K\langle V, V\rangle)\\
&= \int_0^l (\langle \nabla_{\bar{T}} \bar{V}, \nabla_{\bar{T}} \bar{V}\rangle - K\langle \bar{V}, \bar{V}\rangle)\\
&= \int_0^l (\langle \nabla_{\bar{T}} \bar{V}, \nabla_{\bar{T}} \bar{V}\rangle - \langle \bar{R}(\bar{T},\bar{V})\bar{V}, \bar{T}\rangle)\\
&\geq \int_0^l K(\cos^2(t\sqrt{K}) - \sin^2(t\sqrt{K})).
\end{aligned}$$

Since $l < \frac{\pi}{2\sqrt{K}}$, the last term is positive, which suffices to complete the proof. ☐

PROOF OF THEOREM 5.14. The existence of $r(p)$ is an easy consequence of the first statement. Let $q, q' \in \bar{B}_r(p)$. Since

$$\rho(q,q') \leq \rho(q,p) + \rho(p,q') \leq 2r < \mathfrak{I},$$

it follows easily that there is a unique minimal geodesic γ from q to q' which varies continuously with q, q'. Choose $\epsilon > 0$ such that $r + \epsilon < \frac{1}{2}\min(\pi/\sqrt{K}, \mathfrak{I})$, and set

$$V_{r+\epsilon} = \{(q,q') \in \bar{B}_r(p) \times \bar{B}_r(p) | \gamma \subset B_{r+\epsilon}(p)\}.$$

Using the fact that γ varies continuously with q, q' we see that $V_{r+\epsilon}$, which is obviously nonempty, is relatively open. Moreover by Lemma 5.15, if $(q, q') \in V_{r+\epsilon}$ then $\mathrm{int}(\gamma) \subset B_r(p)$. Otherwise $\rho(\gamma(t), p)$ would have an interior maximum. Since γ varies continuously with (q, q') it follows that if $(q, q') \in \bar{V}_{r+\epsilon}$, then $\gamma \subset \bar{B}_r(p) \subset \bar{B}_{r+\epsilon}(p)$. Therefore $\bar{V}_{r+\epsilon} \subset V_{r+\epsilon}$ which is to say that $V_{r+\epsilon}$ is closed. Thus $V_{r+\epsilon}$ is nonempty, relatively open and closed. By connectedness, $V_{r+\epsilon} = \bar{B}_r(p) \times \bar{B}_r(p)$, and as we have seen, the theorem holds for all pairs of points in $V_{r+\epsilon}$. □

CHAPTER 6

The Sphere Theorem and its Generalizations

In this chapter we will show that if M is simply connected and the sectional curvature of M satisfies $1 \geq K_M > \frac{1}{4}$, then M is homeomorphic to a sphere. This result is known as the Sphere Theorem. The symmetric spaces of positive curvature are known to admit metrics such that $1 \geq K_M \geq \frac{1}{4}$; see Example 3.38. In fact, we will prove that any Riemannian manifold with $1 \geq K_M \geq \frac{1}{4}$, which is not a sphere is *isometric* to one of these spaces.

The Sphere Theorem was first proved by Rauch [1951], in 1954, under the assumption $1 \geq K_M \geq \delta \approx \frac{3}{4}$.

Previously, by the use of Hodge theory, Bochner and Yano [1953] had given curvature conditions under which M must be a homology sphere. Rauch's proof used only the Rauch Comparison Theorem. Later, by bringing in the injectivity radius, Klingenberg [1959] simplified the argument and proved the theorem for smaller δ. Then Berger [1960] and Klingenberg [1961] proved it for $\delta = \frac{1}{4}$, which, as we have asserted, is the best possible result. Finally, Berger [1961] proved the rigidity theorem for the case $1 \geq K_M \geq \frac{1}{4}$.

THEOREM 6.1 (The Sphere Theorem). *If M is a complete, simply connected n-dimensional manifold with $1 \geq K_M > \frac{1}{4}$, then M is homeomorphic to the n-sphere S^n.*

To prove this theorem we will need several lemmas. The idea of the proof is to exhibit M as the union of two imbedded balls joined along their common boundary.

We begin with the following general lemma, which is of interest in its own right and useful in other contexts; see Lemma 8.17, Theorem 8.18.

LEMMA 6.2 (Berger). *Let M be any compact Riemannian manifold. Fix $p \in M$ and choose $q \in M$ so that $\rho(p, q)$ is maximal. Then for any vector $v \in T_q$, there exists a minimal geodesic γ from q to p such that $\sphericalangle(\gamma'(0), v) \leq \frac{1}{2}\pi$.*

PROOF. Let σ be the geodesic from q such that $\sigma'(0) = v$. Pick a sequence $\{q_i\}$ on σ such that $q_i \to q$ and let γ_i be a minimal geodesic from q_i to p. If there exists a subsequence γ_j such that $\sphericalangle(\gamma_j', \sigma') \leq \frac{1}{2}\pi$, then the lemma is proved, for any such subsequence must have as a limit point some minimal geodesic γ, and by continuity $\sphericalangle(\gamma'(0), v) \leq \frac{1}{2}\pi$; compare the proof of Lemma 5.2.

Therefore we can assume that there exists some $t_0 > 0$ such that for all $t \in (0, t_0]$ and for any minimal geodesic γ_t from p to $\sigma(t)$ we have $\sphericalangle(\gamma_t', \sigma') > \frac{1}{2}\pi$. Now consider a variation $\{c_s\}$ of γ_t such that c_s goes from p to $\sigma(t-s)$. Using the first variation of arc length (1.5) we find

$$\frac{\mathrm{d}}{\mathrm{d}s}(L[c_s])\bigg|_{s=0} = \langle \gamma_t', \sigma' \rangle < 0.$$

Therefore as t goes from t_0 to 0, the distance from p to $\sigma(t)$ must be strictly decreasing. But this contradicts the fact that $q = \sigma(0)$ is at maximum distance from p. The lemma follows. (See Fig. 6.1.) □

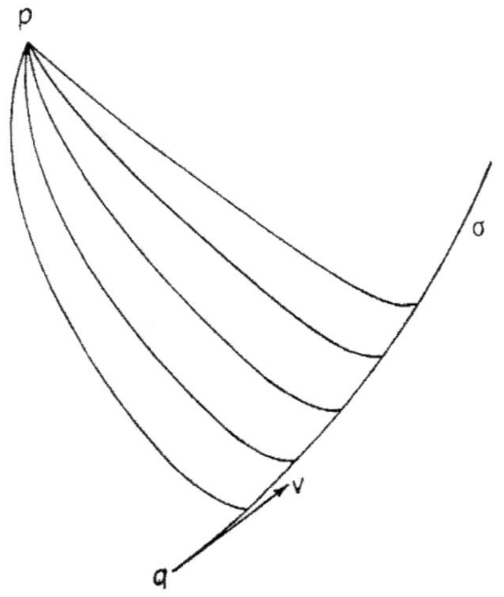

Fig. 6.1.

LEMMA 6.3. *Let $K_M \geq H > 0$ and suppose $d(M) > \frac{\pi}{2\sqrt{H}}$. Then for p,q as above*

$$\overline{B}_{\frac{\pi}{2\sqrt{H}}}(p) \cup \overline{B}_{\frac{\pi}{2\sqrt{H}}}(q) = M.$$

PROOF. Let $K_M \geq H$, and let $(\gamma_1, \gamma_2, \alpha)$ determine a hinge with $\alpha \leq \frac{1}{2}\pi$. Suppose $L[\gamma_i] = l_i > \frac{\pi}{2\sqrt{H}}$. The we claim

$$\rho(\gamma_1(0), \gamma_2(l_2)) < \frac{\pi}{2\sqrt{H}}.$$

In fact, by Toponogov's Theorem, it suffices to check this when M is the sphere of constant curvature H. Using step (1) of the proof of that theorem, we can assume that $\alpha = \frac{1}{2}\pi$. Let θ denote the segment of γ_2 from $\gamma_2(l_2)$ to $\gamma_2(\pi/\sqrt{H})$, and let τ denote the segment of $-\gamma_1$ from $\gamma_1(0)$ to $\gamma_2(\pi/\sqrt{H})$. Then τ and θ meet at the point antipodal to $\gamma_2(0)$ and

$$\sphericalangle(-\theta'(\pi/\sqrt{H} - l_2), -\tau'(\pi/\sqrt{H} - l_1)) = \frac{1}{2}\pi.$$

Since $L[\theta], L[\tau] < \frac{\pi}{2\sqrt{H}}$, it follows that

$$\rho(\theta(0), \tau(0)) = \rho(\gamma_1(0), \gamma_2(l_2)) < \frac{\pi}{2\sqrt{H}}.$$

(See Fig. 6.2.)

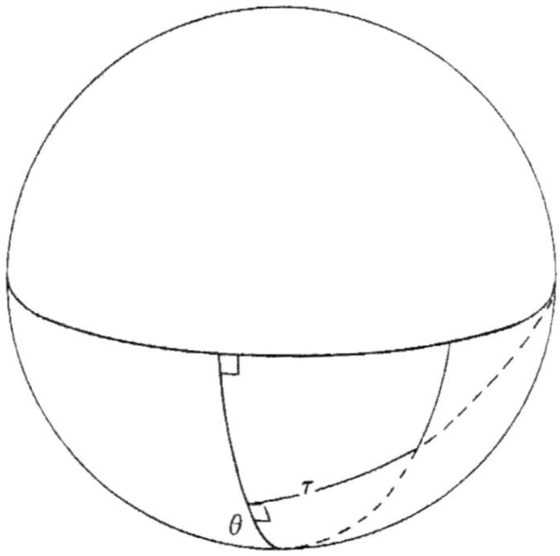

Fig. 6.2.

Now let q be at maximum distance from p and assume $\rho(p, x) > \frac{\pi}{2\sqrt{H}}$. Let γ_2 be a minimal geodesic from p to x and by Lemma 6.2 choose a minimal geodesic γ_1 from q to p such that

$$\sphericalangle(-\gamma_1'(l_1), \gamma_2'(0)) \leq \frac{1}{2}\pi.$$

By the above, $\rho(q, x) < \frac{\pi}{2\sqrt{H}}$. □

REMARK. Tsukamoto [1961] proves this lemma using the Rauch theorem instead of Toponogov's theorem.

LEMMA 6.4. *Suppose $1 \geq K_M \geq H > \frac{1}{4}$, and let p,q be at maximum distance. Then for every geodesic γ from p there exists a unique point $r \in \gamma$ such that*

$$\rho(p, r) = \rho(q, r) \leq \frac{\pi}{2\sqrt{H}}.$$

PROOF. By Theorems 5.9 and 5.10, $i(M) \geq \pi > \frac{\pi}{2\sqrt{H}}$. Therefore we have $\rho(p, \gamma(t)) = t$ for $0 \leq t \leq \pi$. By Lemma 6.3, $\rho(q, \gamma(\pi)) \leq \frac{\pi}{2\sqrt{H}}$. Hence the continuous function $\rho(q, \gamma(t)) - \rho(p, \gamma(t))$ is positive at $t = 0$ and negative at $t = \pi$. Existence of r follows from the Intermediate-Value Theorem. For uniqueness, assume there exist r_1, r_2 on γ each one equidistant from p and q. Assume $\rho(p, r_1) < \rho(p, r_2)$. Then

$$\rho(q, r_2) = \rho(p, r_2) = \rho(p, r_1) + \rho(r_1, r_2) = \rho(q, r_1) + \rho(r_1, r_2).$$

So $\rho(q, r_2) = \rho(q, r_1) + \rho(r_1, r_2)$. If σ is a minimal geodesic from r_1 to q, then the segment of $-\gamma$ from r_2 to r_1 must fit together with σ to form a smooth geodesic to q. Then σ must coincide with the segment of $-\gamma$ from r_1 to p. This implies $q = p$, which is a contradiction. $\square$

PROOF OF THEOREM 6.1. Suppose $1 \geq K_M \geq H > \frac{1}{4}$, and let $p, q \in M$ be at maximal distance. Let S^n denote the unit n-sphere, $\bar{p}, \bar{q} \in S^n$ a pair of antipodal points and $I : S^n_{\bar{p}} \to M^n_p$ an isometry. For each nonzero vector $v \in M^n_p$ define $f(v) = t_0 v$ by letting $\exp f(v)$ be the point along the geodesic $t \to \exp_p tv$ which is equidistant from p and q. The existence of

$$f(v) \leq \frac{\pi}{2\sqrt{H}} < \pi \leq i(M)$$

follows from Lemma 6.4. Thus we can define $h : S^n \to M^n$ by

$$h(x) = \begin{cases} p, & x = \bar{p}, \\ \exp_p\Big(\frac{\rho(x,\bar{p})}{\frac{1}{2}\pi}(f \circ I \circ \exp_p^{-1}(x))\Big), & x \in \overline{B}_{\pi/2}(\bar{p}) - \bar{p}, \\ \exp_q\Big(\frac{\rho(x,\bar{q})}{\frac{1}{2}\pi}(\exp_q^{-1} \circ \exp_p \circ f \circ I \circ \exp_p^{-1}(x))\Big), & x \in \overline{B}_{\pi/2}(\bar{q}) - \bar{q}, \\ q, & x = \bar{q}. \end{cases}$$

We claim:

(1) *h is continuous:* Inspection shows that it suffices to check that $v \to ||f(v)||$ continuous on the unit sphere in M_p. This follows easily from the uniqueness part of Lemma 6.4 and the continuity of the distance function.

(2) *h is injective:* Since $||f|| < i(M)$, it follows that $h|\overline{B}_{\pi/2}(\bar{p})$ is injective as is $h|\overline{B}_{\pi/2}(\bar{q})$. Thus it suffices to show

$$h(B_{\pi/2}(\bar{p}) \cap B_{\pi/2}(\bar{q})) = \emptyset.$$

Suppose for example $x \in h(B_{\pi/2}(\bar{p}))$. Then $x = \gamma(t)$ with $\gamma(0) = p$ and $\rho(x, p) < ||f(\gamma'(0))||$. This uniqueness statement of Lemma 6.4 together with the Intermediate-Value Theorem gives $\rho(x, p) < \rho(x, q)$. By the same argument, if $x \in h(B_{\pi/2}(\bar{q}))$, then $\rho(x, q) < \rho(x, p)$. Therefore $x \in h(B_{\pi/2}(\bar{p}))$ implies $x \notin h(B_{\pi/2}(\bar{q}))$

(3) *h is surjective:* If $\rho(x, p) \leq \rho(x, q)$, let γ be a minimal normal geodesic from p to x and define t by the equation $x = \gamma(t)$. By Lemma 6.4 and the argument above there exists $t_0 \geq t$ such that $t_0 = ||f(v)||$. Then clearly $x \in \overline{B}_{\pi/2}(\bar{p})$. By a symmetric argument, if $\rho(x, q) \leq \rho(x, p)$, then $x \in \overline{B}_{\pi/2}(\bar{q})$.

Since h is an injective, surjective, continuous map between compact manifolds, it follows that h is a homeomorphism. $\square$

THEOREM 6.5 (Maximal Diameter Theorem (Toponogov)). *Let M be a complete n-manifold such that $K_M \geq H > 0$ and suppose that the diameter of M is equal to $\pi/\sqrt{H}$. Then M is isometric to $S^n_{1/\sqrt{H}}$, the sphere of curvature H.*

PROOF. Pick $p, q \in M$ such that $\rho(p,q) = \pi/\sqrt{H}$. We shall show that all normal geodesics from p reach q at time $\pi/\sqrt{H}$ and are therefore minimal to q.

Let

$$\gamma_1 : [0, t_0] \to M, \qquad 0 < t_0 < \pi/\sqrt{H},$$

be any geodesic segment such that $\gamma_1(t_0) = p$, and let

$$\gamma_2 : [0, \pi/\sqrt{H}] \to M$$

be any minimal geodesic from p to q. Applying (B) of Toponogov's Theorem gives $\rho(q, \gamma_1(0)) = \frac{\pi}{\sqrt{H}} - t_0$. Thus if σ is a minimal geodesic from q to $\gamma_1(0)$, then

$$L[\sigma \cup \gamma_1] = \pi/\sqrt{H}.$$

So $\sigma \cup \gamma_1$ must form a smooth minimal geodesic from p to q.

It follows that $\exp_p | B_{\pi/\sqrt{H}}(0) \subset M_p$ is nonsingular. Moreover, if γ is any geodesic from p to q and V is any Jacobi field along γ which vanishes at $\gamma(0)$, then V must also vanish at $\gamma(\pi/\sqrt{H})$. The argument of Proposition 4.13 shows that all sectional curvatures of sections spanned by γ', V must be equal to H. Lemma 1.41 implies that $B_{\pi/\sqrt{H}}(p)$ is isometric to $B_{\pi/\sqrt{H}}(\bar{p})$, where $\bar{p} \in S^n_{1/\sqrt{H}}$. By continuity all plane sections of M have curvature H, so M has constant curvature. By Theorem 1.43 the universal covering space of M is isometric to S^n. On the other hand if $I : (S^n_{1/\sqrt{H}})_{\bar{p}} \to M^n_p$ is any isometry, then $f = \exp_p \circ I \circ \exp_{\bar{p}}^{-1}$ obviously extends to a homeomorphism between $S^n_{1/\sqrt{H}}$ and M^n. So M^n must itself be isometric to $S^n_{1/\sqrt{H}}$. Alternatively it is not difficult to see directly that f extends to an isometry. □

In the course of the following theorem we will be using implicitly the estimate $i(M) \geq \pi$, if dim M is odd, M is simply connected and $1 \geq K_M \geq \frac{1}{4}$. We have proved this only in the case $1 \geq K_M > \frac{1}{4}$; see the remarks preceding Theorem 5.10, and Cheeger and Gromoll [1972].

THEOREM 6.6 (Minimal Diameter Theorem (Berger)). *Let M be a complete simply connected manifold such that $1 \geq K_M \geq \frac{1}{4}$.*

(1) If $d(M) > \pi$, M is homeomorphic to S^n.

(2) If $d(M) = \pi$, M is isometric to a symmetric space.

The proof of this theorem is quite long and intricate. In order to prove (1) we shall require three preliminary lemmas.

LEMMA 6.7. *Fix $p \in M$ and assume that $1 \geq K(\sigma) \geq H > 0$ for all plane sections σ. Let $x, y, z \in M_p$ be linearly independent unit vectors. Assume that x is not perpendicular to both y and z. Let $K(x,y)$ denote the curvature of the plane spanned by x,y and suppose $K(x, y) = K(x, z) = H$. Then $K(y, z) < 1$.*

PROOF. It follows from the usual symmetry properties of the curvature tensor that $u \to R(u,x)x$ is s symmetric linear transformation. The smallest nonzero eigenvalue of this transformation is at least H. Therefore if

$$\langle u, x\rangle = \langle v, x\rangle = \langle u, v\rangle = 0$$

and $K(u,x) = H$, then

$$R(u,x)x = Hu, \qquad \langle R(u,x)x, v\rangle = 0.$$

Now given x, y, z as above, choose u, v such that x, u, v are orthonormal and

$$y = ax + bu, \qquad z = cx + du + ev.$$

Then using

$$R(u,x)x = Hu, \quad R(v,x)x = Hv, \quad R(x,u)u = Hx, \quad R(x,v)v = Hx,$$

one computes directly that

$$\begin{aligned} K(y,z) &= \frac{(ad-bc)^2 H + e^2(a^2 H + b^2 K(u,v))}{||y||^2||z||^2 - \langle y, z\rangle^2} \\ &= \frac{((ad-bc)^2 + (a^2+b^2)e^2)H + e^2 b^2 (K(u,v) - H)}{1-(ac+bd)^2}. \end{aligned}$$

Using

$$||y||^2 = a^2 + b^2 = ||z||^2 = c^2 + d^2 + e^2 = 1,$$

one computes easily that

$$(ad-bc)^2 + e^2 = 1 - (ac+bd)^2.$$

Therefore,

$$K(y,z) = H + \frac{e^2 b^2 (K(u,v) - H)}{(ad-bc)^2 + e^2} \leq 1.$$

Equality holds if and only if $K(u,v) = 1, b = 1, a = 0, c = 0$, which completes the proof. $\square$

Now we consider a special case of Lemma 6.2.

LEMMA 6.8. *Let $p, q \in M$ be chosen so that $\rho(p,q)$ is maximal and fix $v \in M_q$. Assume there is no minimal geodesic γ from q to p such that $\measuredangle(\gamma', v) < \frac{1}{2}\pi$. Then there exist at least two minimal geodesics λ, λ_0 from q to p such that*

$$\measuredangle(\lambda', v) = \measuredangle(\lambda_0', v) = \frac{1}{2}\pi.$$

PROOF. By Lemma 6.2, we know there must be one such geodesic; call it λ. Now consider the vector $v - t\lambda'$ $(t > 0)$. By Lemma 6.2, there exists a minimal geodesic λ_t from q to p such that

$$\langle \lambda_t', v - t\lambda'\rangle \geq 0.$$

But $\langle \lambda_t', v\rangle \leq 0$. Hence $\langle \lambda_t', \lambda'\rangle \leq 0$. As $t \to 0$, the collection of geodesics $\{\lambda_t\}$ has at least one limit point; let it be λ_0. Then $\langle \lambda_0', \lambda'\rangle \leq 0$, so λ and λ_0 are distinct and $\langle \lambda', v\rangle = \langle \lambda_0', v\rangle = 0$. $\square$

LEMMA 6.9. *Let* $1 \geq K_M$ *and* $i(M) \geq \pi$. *Let* $\tau_1, \tau_2 : [0, \pi] \to M$ *be distinct geodesics such that* $\tau_1(0) = \tau_2(0)$, $\tau_1(\pi) = \tau_2(\pi)$ *and* $\sphericalangle(\tau_1'(0), \tau_2'(0)) \neq \pi$. *Then for all unit vectors*

$$V = \alpha\tau_1'(0) + \beta\tau_2'(0),$$

with $\alpha, \beta \geq 0$, *we have*

$$\exp_{\tau_1(0)} \pi V = \tau_1(\pi).$$

The union of the geodesics

$$\exp_{\tau_1(0)} tV, \qquad 0 \leq t \leq \pi,$$

is an imbedded totally geodesic piece of surface of curvature 1. In particular, the plane section spanned by $\tau_1'(0), \tau_2'(0)$ *has curvature 1.*

PROOF. For $t \leq \frac{1}{2}\pi$, consider the path $-\tau_2 \cup \tau_1$ from $\tau_2(\frac{1}{2}\pi)$ to $\tau_1(t)$. Since this path is not smooth, it has length $< \pi$. Similarly, for $\pi \geq t > \frac{1}{2}\pi$,

$$\rho(\tau_1(t), \tau_2(\frac{1}{2}\pi)) < \pi.$$

Therefore we may apply Corollary 1.35, with $m = \tau_2(\frac{1}{2}\pi)$, $c = \tau_1$ and M_0 the sphere of curvature 1.

Let $\mathfrak{L}$ denote the lune bounded by $\exp_{m_0} \circ I \circ \exp_m^{-1}(\tau_1) = c_0$ and $\exp_{m_0} \circ I \circ \exp_m^{-1}(\tau_2)$. Since $L[c] = L[c_0] = \pi$ it follows from the remarks after the proof of Theorem 1.34, and from the proof of Corollary 1.35, that $\exp_m \circ I^{-1} \circ \exp_{m_0}^{-1} | \mathfrak{L}$ is an immersion whose differential is an isometry at each point of $\mathfrak{L}$. The image of a geodesic in $\mathfrak{L}$ which runs from $c_0(0)$ to $c_0(\pi)$ is thus a curve of length π. Hence since $\rho(c(0), c(\pi)) = \pi$, it is a geodesic in M. It follows that $\exp_m \circ I^{-1} \circ \exp_{m_0}^{-1}(\mathfrak{L})$ is totally geodesic of curvature 1. This suffices to complete the proof. □

PROOF OF THEOREM 6.6(1). By Theorem 6.5, we can assume that the diameter $d(M)$ satisfies $\pi < d(M) = l < 2\pi$. Pick $p, q \in M$ such that $\rho(p, q) = d(M)$. Suppose M can be written as the union of open balls $M = B_\pi(p) \cup B_\pi(q)$. Then

$$M = \overline{B}_r(p) \cup \overline{B}_r(q)$$

for some $r < \pi$, and we may proceed exactly as in the Sphere Theorem. Therefore assume that there is a point x such that $\rho(p, x) \geq \pi$ and $\rho(q, x) \geq \pi$. Let σ be a minimal geodesic from p to x and choose $\gamma_1 : [0, l] \to M$ to be minimal from q to p such that $\sphericalangle(-\gamma_1'(l), \sigma'(0)) \leq \frac{1}{2}\pi$. Let $(\bar{\gamma}, \bar{\sigma}, \bar{\alpha}$ determine a hinge on the sphere of curvature $\frac{1}{4}$ with

$$L[\bar{\gamma}] = L[\gamma_1], \qquad L[\bar{\sigma}] = L[\sigma]$$

and

$$\bar{\alpha} = \sphericalangle(-\gamma_1'(l), \sigma'(0)).$$

Then reasoning as in the proof of Lemma 6.3, we see that $L[\sigma] = \pi, \bar{\alpha} = \frac{1}{2}\pi$ and

$$\rho(\gamma_1(0), \sigma(\pi)) = \rho(\bar{\gamma}_1(0), \bar{\sigma}(\pi)) = \pi.$$

Since we have equality, Corollary 2.3, the rigid version of Toponogov's Theorem, applies. Thus if $\bar{\tau}_1$ is the minimal geodesic from $\bar{\sigma}(\pi)$ to $\bar{\gamma}_1(0)$, we obtain a corresponding geodesic τ_1 from $\sigma(\pi)$ to $\gamma_1(0) = q$ such that $L[\tau_1] = \pi$ and γ_1, σ, τ_1 span an embedded totally geodesic piece of surface having curvature $\frac{1}{4}$. In particular

(1) $\pi > \sphericalangle(-\sigma'(\pi), \tau_1'(0)) = \frac{1}{2}l > \frac{1}{2}\pi$,

(2) $\tau_1'(0)$ is in the plane spanned by $\sigma'(0)$ and the image of $\gamma_1'(0)$ under parallel translation.

(3) The curvature of the plane section spanned by $\tau_1'(0), \sigma'(\pi)$ is equal to $\frac{1}{4}$.

Since we were able to conclude that for arbitrary γ_1, $\sphericalangle(-\gamma_1'(l), \sigma'(0)) \le \frac{1}{2}\pi$ implies $\sphericalangle(-\gamma_1'(l), \sigma'(0)) = \frac{1}{2}\pi$, it follows from Lemma 6.8 that there exists a geodesic γ_2 from q to p which is distinct from γ_1 and such that $\sphericalangle(-\gamma_2'(l), \sigma'(0)) = \frac{1}{2}\pi$. The above argument shows the existence of τ_2 from x to q satisfying the same three conditions as τ_1. Conditions (1) and (2) imply that $\tau_1'(0) \neq \tau_2'(0)$ and that

$$\sphericalangle(-\sigma'(\pi), \tau_1'(0)) \neq \frac{1}{2}\pi, \qquad \sphericalangle(-\sigma'(\pi), \tau_2'(0)) \neq \frac{1}{2}\pi.$$

Lemma 6.9 implies that the curvature of the section spanned by $\tau_1'(0), \tau_2'(0)$ is equal to 1. But if $\tau_2'(0)$ lies in the plane spanned by $\sigma'(\pi), \tau_1'(0)$, this contradicts (3) above. If $\sigma'(\pi), \tau_1'(0), \tau_2'(0)$ are linearly independent, then applying Lemma 6.7, we again contradict (3). This proves 6.6(1). □

The proof of Theorem 6.6(2) will also require three lemmas.

LEMMA 6.10. *Let $K_M \ge H$, and $\tau : [0, l] \to M$ be a geodesic.*
(1) If there exists $p \in M$ such that

$$\rho(\tau(0), p) \ge \frac{\pi}{2\sqrt{H}}, \qquad \rho(\tau(l), p) > \frac{\pi}{2\sqrt{H}}, \qquad l \le \pi/\sqrt{H}.$$

then the function $\rho(\tau(t), p)$ does not have a weak minimum for $0 < t < l$.
(2) If there exists $p \in M$ such that

$$\rho(\tau(0), p) = \rho(\tau(l), p) = \frac{\pi}{2\sqrt{H}} \qquad l \le \pi/\sqrt{H},$$

then the function $\rho(\tau(t), p)$ does not have a weak minimum unless $\rho(\tau(t), p) \equiv \frac{\pi}{2\sqrt{H}}$.

PROOF. (1) Let $\rho(\tau(t), p)$ have a weak minimum at t_0. Let $\sigma : [0, l_1] \to M$ be a minimal geodesic from p to $\tau(t)$. Then $\langle \sigma'(l_1), \tau'(t_0) \rangle = 0$. Thus σ, and $\tau|[t_0, l]$ determine a *right hinge*. If $L[\sigma] \le \frac{\pi}{2\sqrt{H}}$ and $t_0 < \frac{\pi}{2\sqrt{H}}$, then using Toponogov's Theorem as in Lemma 6.3 we find that

$$\rho(\tau(0), p) < \frac{\pi}{2\sqrt{H}}$$

(see Fig. 6.2). Similarly, $l - t \leq \frac{\pi}{2\sqrt{H}}$ implies

$$\rho(\tau(l), p) \leq \frac{\pi}{2\sqrt{H}}$$

in this case. In the same way $L[\sigma] \leq \frac{\pi}{2\sqrt{H}}$ implies that $\rho(\tau(0), p)$ and $\rho(\tau(l), p)$ are less than $\rho(\tau(t_0), p) = L[\sigma]$.

(2) The argument is the same as above. □

Lemma 6.10 may be paraphrased as saying that if $K_M \geq H$, then the complement of a ball of radius $\frac{\pi}{2\sqrt{H}}$ is convex.

LEMMA 6.11. *Let $1 \geq K_M \geq \frac{1}{4}$ and $\tau : [0, l] \to M$ be such that $p(\tau(t), p) \equiv \pi$ for $0 \leq t \leq l$. Then for every minimal geodesic $\gamma : [0, \pi] \to M$ such that $\gamma(0) = p$ and $\gamma(\pi) = \tau(0)$ we have*

$$\langle \gamma'(\pi), \tau'(0) \rangle = 0.$$

PROOF. Let σ_ϵ be a minmal geodesic from p to $\tau(\epsilon)$. Then by the first variation formula $\langle \sigma'_\epsilon(\pi), \tau'(\epsilon) \rangle = 0$. Let $\sigma : [0, \pi] \to M$ be a geodesic such that $\sigma'(0)$ is a limt point of $\{\sigma'_\epsilon(0)\}$. Then

$$\sigma(\pi) = \tau(0), \qquad \langle \sigma'(\pi), \tau'(0) \rangle = 0.$$

Let $E(t)$ be the parallel field along σ generated by $\tau'(0)$. Consider the curve

$$h_\epsilon = \exp_{\sigma(t)} \epsilon \sin(\frac{1}{2}t) E(t).$$

By Rauch II, comparing with a sphere of curvature $\frac{1}{4}$, $L[h_\epsilon] \leq \pi$, and h_ϵ is a geodesic for each ϵ. Therefore there is a Jacobi field J along σ such that

$$J(0) = 0, \qquad J(\pi) = E(\pi) = \tau'(0).$$

Let $\tilde{c} : (-\epsilon, \epsilon) \to M_p$ be such that

$$||\tilde{c}(s)|| \equiv \pi, \qquad \tilde{c}(0) = \pi\sigma'(0), \qquad \mathrm{dexp}_p \tilde{c}'(0) = J(\pi) = \tau'(0).$$

Set

$$\exp_p \tilde{c}(s) = c(s).$$

Since $i(M) \geq \pi$, we have $\rho(c(s), p) = \pi$. Therefore by the first variation formula, for any minimal geodesic γ from p to $c(0) = \tau(0)$ we have

$$\langle \gamma'(0), c'(0) \rangle = \langle \gamma'(0), \tau'(0) \rangle = 0.$$

□

LEMMA 6.12. *Let $1 \geq K_M \geq \frac{1}{4}$, $d(M) = \pi$, and let $\gamma : [0, \pi] \to M$ be a normal geodesic. Let $v \in M_{\gamma(\pi)}$ be such that there exists some Jacobi field J along γ with $J(0) = 0$ and $J(\pi) = v$. Then if $E(t)$ is the parallel field along γ generated by v, $\sin \frac{1}{2}tE(t)$ is a Jacobi field along γ. In particular, the curvature of the plane section spanned by $\gamma'(t)$ and $E(t)$ satisfies*

$$K(\gamma'(t), E(t)) \equiv \frac{1}{4}.$$

PROOF. Let $c : (-\epsilon, \epsilon) \to M$ constructed as in the proof of the previous lemma such that

$$c(0) = \tau(0), \qquad c'(0) = v.$$

Let $\tau_\epsilon : [0, l_\epsilon] \to M$ be the unique minimal geodesic from $c(0)$ to $c(\epsilon)$. By Lemma 6.10(2) and the assumption that $d(M) = \pi$, it follows that $\rho(\tau_\epsilon(s), p) \equiv \pi$ for $0 \le s \le l_\epsilon$. By Lemma 6.11 it follows that $\langle \tau_\epsilon'(0), \gamma'(\pi) \rangle = 0$ and by the proof of that lemma, $\sin \frac{1}{2} t E_\epsilon(t)$ is a Jacobi field. That $K(\gamma'(t), E(t)) \equiv \frac{1}{4}$ then follows from the Jacobi equation. □

PROOF OF THEOREM 6.6(2). Let $\gamma : [0, \pi] \to M$ be a geodesic. Let $\mathfrak{J}_1$ denote the space of Jacobi fields along γ vanishing at $\gamma(0), \gamma(\pi)$. Since $1 \ge K_M$, the elements of $\mathfrak{J}_1$ are of the form $\epsilon \sin t F(t)$, where $F(t)$ is a parallel field and $K(\gamma'(t), F(t)) \equiv 1$. By Lemma 6.12, the space $\mathfrak{J}$ of Jacobi fields vanishing at $\gamma(0)$ splits up as $\mathfrak{J}_1 \bigoplus \mathfrak{J}_{\frac{1}{4}}$, where the members of $\mathfrak{J}_{\frac{1}{4}}$ are of the form $\sin \frac{1}{2} t E(t)$. Since $\frac{1}{4}$ and 1 are eigenvalues of the symmetric transformation $x \to R(x, \gamma')\gamma'$, we have

$$\langle \mathfrak{J}_{\frac{1}{4}}, \mathfrak{J}_1 \rangle \equiv 0$$

and the decomposition $\mathfrak{J} = \mathfrak{J}_1 \bigoplus \mathfrak{J}_{\frac{1}{4}}$ is unique. Therefore it agrees with the corresponding decomposition for $\gamma|[t_0, \pi + t_0]$ for $0 < t_0 < \pi$. In this way we see that the decomposition $\mathfrak{J} = \mathfrak{J}_1 \bigoplus \mathfrak{J}_{\frac{1}{4}}$ extends for all values of t. Now let $I_{\gamma(0)}$ denote the local geodesic symmetry about $\gamma(0)$ (see Proposition 3.37). If J is a Jacobi field along γ such that $J(0) = 0$, then it is easily seen that

$$\mathrm{d}I_{\gamma(0)}(J(t)) = -J(-t).$$

Thus $\mathrm{d}I_{\gamma(0)}$ preserves $\mathfrak{J}_1, \mathfrak{J}_{\frac{1}{4}}$ and clearly, restricted to these subspaces, it is norm preserving. It follows that I_{γ_0} is a local isometry. Hence, since M is simply connected, by Proposition 3.37, M is a symmetric space. □

It follows from the classification of symmetric spaces that the only simply connected symmetric spaces with positive curvature are S^n, the projective spaces $CP(n), QP(n)$ and the Cayley Plane; see Helgason [1962].

We now give a somewhat weakened generalization of the Sphere Theorem.

THEOREM 6.13 (Homotopy Sphere Theorem). *Let M be a complete simply connected manifold such that $K_M \ge H > 0$, such that $d(M) > \frac{\pi}{2\sqrt{H}}$. Then M is a homotopy sphere.*

PROOF. Fix $p, q \in M$ so that $\rho(p, q) = d(M)$, and let γ be a closed geodesic from p. It follows from Lemma 6.10 that $L[\gamma] > \pi/\sqrt{H}$. By Proposition 4.13 it follows that the index of γ is $\ge n - 1$. After possibly approximating the energy function by one with nondegenerate critical points, from Morse Theory (Theorem 4.10) it follows that $\Omega(p, p)$ the loop space at p is $(n-2)$-connected. Therefore M is $(n-1)$-connected, so it is a homotopy sphere. □

REMARK 6.14. From the Poincaré conjecture (Perelman [2002], [2003]) and the generalized Poincaré conjecture (Smale [1956], Friedman [1982]) it follows that any M which satisfies the hypothesis of the above theorem is a topological sphere.

CHAPTER 7

The Differentiable Sphere Theorem

In this chapter we will refine the Sphere Theorem by showing that there exists a constant δ such that if M is a complete simply connected n-dimensional Riemannian manifold satisfying $1 \geq K_M > \delta$, then M is *diffeomorphic* to the standard sphere S^n. Such an M is called δ-pinched. It is known that for $n \geq 7$ there exist manifolds which are homeomorphic to but not diffeomorphic to the n-sphere with its standard differential structure. However, it is not known that any of these "exotic" spheres actually admit Riemannian metrics of strictly positive sectional curvature. Thus it is conceivable that the best possible value for δ in the differentiable case is again $\frac{1}{4}$. As we have mentioned in Example 3.44, there does exist an exotic sphere which carries at least a metric of nonnegative curvature. On the other hand, by applying a result of Singer and Lichnerowicz, Hitchin [1972] has observed that some exotic spheres do not even admit metrics of positive scalar curvature.

The Differentiable Sphere Theorem was first proved by Gromoll [1966] and Calabi. However, they needed a different constant δ_n for each dimension and $\lim_{n\to\infty} \delta_n = 1$. It was shown recently by Shiohama that δ can be chosen independent of n. The actual value of δ was improved succesively by Sugimoto, Shiohama and Karcher [1971] and Ruh [1971]. The most recent estimate is $\delta \approx 0.80$. In order to avoid nonconceptual complications, we shall content ourselves with showing the existence of *some* δ which is independent of n.

The main point in the proof of the Sphere Theorem of Chapter 6 was to show that if M is simply connected and satisfies $1 \geq K_M \geq H > \frac{1}{4}$, then M may be written as the union of two embedded balls. In order to handle the differentiable case, we will refine this condition as follows. Let $B^n_r \subset \mathbb{R}^n$ denote the ball of radius r.

DEFINITION 7.1. A smooth manifold M^n is called a *twisted sphere* if there are smooth embeddings $h_1, h_2 : B^n_{1+\epsilon} \to M^n$ such that

$$h_1(\bar{B}^n_1) \cup h_2(\bar{B}^n_1) = M^n, \qquad h_1(B^n_1) \cap h_2(B^n_1) = \emptyset.$$

The proof of the Sphere Theorem shows that a twisted sphere is homeomorphic to the standard sphere. Let $S^{n-1} = \partial B^n_1$ or ∂B^n_2. The function $f = h_2^{-1} \circ h_1 | S^{n-1}$ is a diffeomorphism $f : S^{n-1} \to S^{n-1}$. Recall that a diffeomorphism f is said to be *isotopic* to a diffeomorphism g if there exists a smooth 1-parameter family of diffeomorphisms F_t with $F_1 = f$ and $F_0 = g$. A basic fact about twisted spheres is that M^n is diffeomorphic to the standard sphere if $f = h_2^{-1} \circ h_1$ is isotopic to the identity map. Moreover, in order to show that f is isotopic to the identity it suffices to check that for all $y \in S^{n-1}$ and $v \in S^{n-1}$ with $||v|| = 1$,

$$\rho(y, f(y)) \leq \frac{1}{2}\pi, \text{ and } ||v - \mathrm{d}f_y(v)|| < 1. \qquad (*)$$

One might object to the second inequality here because v and $\mathrm{d}f_y(v)$ are in different tangent spaces. We consider S^{n-1} with its standard embedding in $\mathbb{R}^n$ so all tangent spaces are hyperplanes in $\mathbb{R}^n$ and thus such a subtraction is permitted.

Once these results have been established, we observe that if M^n is simply connected and $1 \geq K_M \geq \delta > \frac{1}{4}$, then M^n is a twisted sphere. Further, if δ is sufficiently close to 1, then f satisfies $(*)$. Hence f is isotopic to the identity and M^n is diffeomorphic to S^n.

PROPOSITION 7.2. *Let M^n be a twisted sphere. If f is isotopic to the identity, then M^n is diffeomorphic to the standard sphere S^n.*

PROOF. Let

$$\tilde{h}_1(B^n_{1+\epsilon}) \cup \tilde{h}_2(B^n_{1+\epsilon}) = S^n$$

denote the usual covering of S^n, i.e. $\tilde{h}_2$ is $\tilde{h}_1$ followed by reflection in the equator. Then for $1 - \epsilon < t < 1 + \epsilon$ and $y \in S^{n-1}$, we have

$$\tilde{h}_2^{-1} \circ \tilde{h}_1(ty) = (2 - t)y.$$

It will suffice to find a cover of M^n by embeddings $k_i : B^n_{1+\epsilon} \to M^n$ such that

$$k_2^{-1} \circ k_1(ty) = (2 - t)y.$$

Then we may define a diffeomorphism $H : M^n \to S^n$ by setting

$$H|k_i(B^n_{1+\epsilon}) = \tilde{h}_i \circ \tilde{k}_i^{-1}.$$

In order to produce the k_i, the h_i must be suitably modified. Set

$$V_i = \mathrm{d}h_i\Big(\frac{\partial}{\partial r}\Big), \qquad N_{\epsilon_1} = h_1(B^n_{1+\epsilon_1} - \overline{B}^n_{1-\epsilon_1}).$$

Choose ϵ_1 sufficiently small so that $N_{\epsilon_1} \subset h_2(B^n_{1+\epsilon})$ and such that on N_{ϵ_1} we have

$$V_1(||h_2^{-1}(x)||) < 0, \qquad V_2(||h_1^{-1}(x)||) < 0. \tag{7.3}$$

Let $\alpha : [0,1] \to [0,1]$ satisfy

$$\alpha(0) = 0, \qquad \alpha(1) = 1,$$
$$\alpha'(x) \geq 0,$$
$$\alpha'(0) = 0, \qquad \alpha'(1) = 0.$$

Define a vector field W on N_{ϵ_1} by

$$W(h_1(ty)) = (1 - \alpha(u))V_1 - \alpha(u)V_2,$$

where

$$u = \frac{t}{2\epsilon_1} - \frac{1-\epsilon_1}{2\epsilon_1}.$$

Define vector fields $V_{1,s}$ on $h_1(B^n_{1+\epsilon_1})$ and $V_{2,s}$ on $h_2(B^n_1) \cup N_{\epsilon_1}$ by

$$V_{1,s}(x) = \frac{(1-s)V_1 + sW}{((1-s)V_1 + sW)(||h_1^{-1}(x)||)},$$

$$V_{2,s}(x) = \frac{(1-s)V_2 - sW}{((1-s)V_2 - sW)(||h_2^{-1}(x)||)}.$$

(7.3) guarantees that the denominator above does not vanish. Let $\psi_{i,s,z}$ denote the integral curve of $V_{i,s}$ with $\psi'_{i,s,z}(0) = z$. Define

$$h_{i,s}(x) = \psi_{i,s,z}(t),$$

where t is defined by

$$h_i(x) = h_{i,0}(x) = \psi_{i,0,z}(t).$$

Using $V_{i,s}(||h_i^{-1}(x)||) > 0$, it follows easily that $h_{i,s}(x)$ is an embedding. We have

$$h_{1,s}(\partial \bar{B}^n_1) = h_{2,s}(\partial \bar{B}^n_1).$$

Set

$$h_{2,s}^{-1} \circ h_{1,s}|S^{n-1} = f_s.$$

Then

$$h_{2,1}^{-1} \circ h_{1,1}(ty) = g(y,t)f_1(y),$$

where

$$g : S^{n-1} \times [1-\epsilon_1, 1+\epsilon_1] \to S^{n-1} \times [1-\epsilon_1, 1+\epsilon_1]$$

is a suitable smooth function with $\partial g/\partial t < 0$. (See Fig. 7.1.)

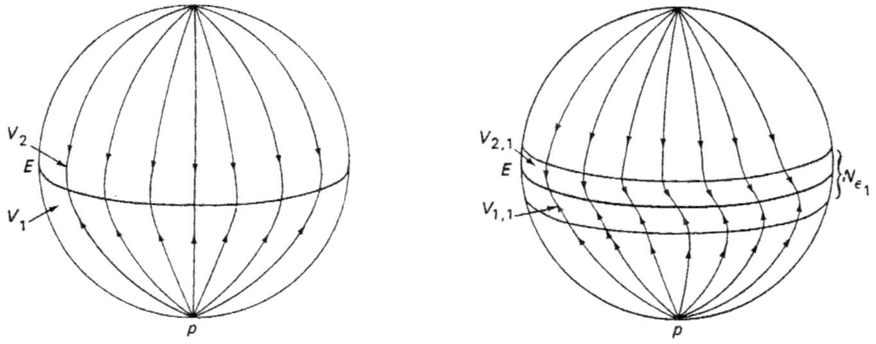

Fig. 7.1.

Let

$$\beta(y,t) : S^{n-1} \times [0, 1+\epsilon_1] \to \mathbb{R}^+$$

be a smooth function such that

$$\partial\beta/\partial t > 0,$$

$$\beta(y,t) = \begin{cases} t, & 0 \le t \le \frac{1}{2}(1-\epsilon_1), \\ g(y, 2-t), & 1-\epsilon_1 \le t \le 1+\epsilon_1. \end{cases}$$

Set

$$\hat{h}_1 = h_{1,1}, \qquad \hat{h}_2(ty) = h_{2,1}(\beta(y,t)y), \qquad \hat{f} = f_1.$$

Then

$$\hat{h}_2^{-1} \circ \hat{h}_1(ty) = (2-t)\hat{f}(y).$$

Finally, let $0 < r_1 < r_2 < 1 - \epsilon_1$, and let $F_t : S^{n-1} \times [0,1] \to S^{n-1}$ be an isotopy with $F_0 = 1$ and $F_1 = \hat{f}$. Set $k_1 = h_1$ and define k_2 by

$$k_2(ty) = \begin{cases} \hat{h}_2(ty), & 0 \le t \le r_1, \\ \hat{h}_2(tF_{\alpha(u_0)}(y)), & r_1 \le t \le r_2, \\ \hat{h}_2(t\hat{f}(y)), & r_2 \le t \le 1+\epsilon_1, \end{cases}$$

where

$$u_0 = \frac{t}{r_2 - r_1} - \frac{r_1}{r_2 - r_1}$$

and α is the function defined previously. (See Fig. 7.2.) Then

$$k_2^{-1} \circ k_1(ty) = k_2^{-1} \circ \hat{h}_1(ty) = k_2^{-1} \circ \hat{h}_2((2-t)\hat{f}(y)) = (2-t)y.$$

This suffices to complete the proof. □

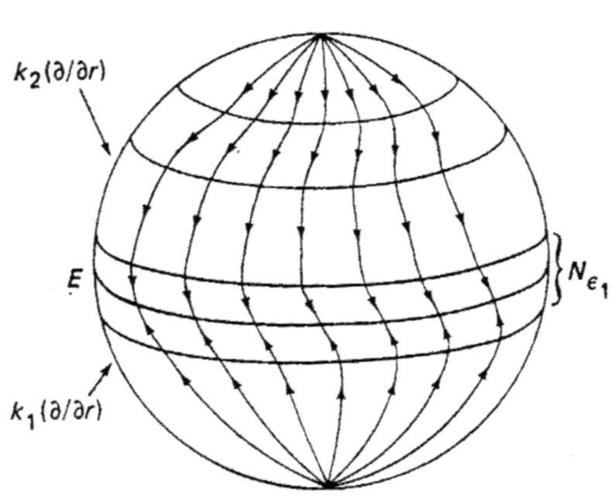

Fig. 7.2.

The following proposition is not as sharp as possible; see Sugimoto et al. [1971].

PROPOSITION 7.4. *Let $f : S^{n-1} \to S^{n-1}$ be a diffeomorphism of the unit sphere. If for all $y \in S^{n-1}, v \in S_y^{n-1}, ||v|| = 1$, we have*

$$\rho(y, f(y)) \leq \frac{1}{2}\pi, \qquad ||v - \mathrm{d}f_y(v)|| \leq 1,$$

then f is isotopic to the identity map.

PROOF. Since in particular, y and $f(y)$ are not antipodal, the map

$$f_t(y) = \exp_y(t \exp_y^{-1} f(y))$$

is well defined and smooth. Thus we need only check that f_t is a diffeomorphism for each t. To do this we just have to check that $\mathrm{d}f_t$ is nonsingular. Let $\phi(s)$ be a curve such that

$$\phi(0) = u, \qquad \phi'(0) = v.$$

Since $t \to f_t(\phi(s))$ is a geodesic, it follows that

$$\mathrm{d}f_t(v) = \frac{\mathrm{d}}{\mathrm{d}s} f_t(\phi(s))$$

is a Jacobi field $Z(t)$ along $f_t(u)$. Set $\beta = \rho(y, f(y))$. If $\beta = 0$, then

$$Z(t) = t\mathrm{d}f(v) + (1-t)v.$$

Clearly, $Z(t) \neq 0$ unless $\mathrm{d}f_y(v) = -kv$ for some $k \geq 0$. But then, if $||v|| = 1$, we have $||v - \mathrm{d}f_y(v)|| \geq 1$. So assume that $\beta \neq 0$. Then by Proposition 1.16 and its sequel,

$$Z(t) = (at + b)T(t) + c \sin \beta t E + d \cos \beta t F,$$

where

$$\frac{\mathrm{d}}{\mathrm{d}t}(f_t(y)) = T(t)$$

and E, F are perpendicular to $T(t)$. Now if $Z(t_0) = 0$, clearly $E = F$. Since for $0 \leq t \leq 1$ and $0 \leq \beta \leq \frac{1}{2}\pi$ we have $\sin \beta t, \cos \beta t \geq 0$, it follows that c and d have opposite signs, as do b and $a + b$. Since

$$Z(0) - Z(1) = bT(0) - (a+b)T(1) + (d - c \sin \beta)E$$

and $\langle T(0), T(1) \rangle \geq 0$, clearly

$$||v - \mathrm{d}f_y(v)|| = ||Z(0) - Z(1)|| \geq ||Z(0)|| = 1,$$

which is a contradiction. Thus $Z(t)$ does not vanish. □

In order to verify the hypothesis of Proposition 7.4 it will be helpful to have the following result.

LEMMA 7.5. *There is a function $\mu(\epsilon) \geq 0$ satisfying $\lim\limits_{\epsilon \to 0} \mu(\epsilon) = 0$ with the following property. Let $f : S^{n-1} \to S^{n-1}$ be a diffeomorphism such that for all v with $||v|| = 1$:*

(1) $1 - \epsilon \leq ||\mathrm{d}f_y(v)|| \leq 1/(1-\epsilon)$;

(2) there exists $y_0 \in S^{n-1}$ such that $\rho(y_0, f(y_0)) < \epsilon$, and for all y_1 with $\rho(y_0, y_1) = \frac{1}{2}\pi$ we have $\rho(y_1, f(y_1)) < \epsilon$.

Then for all y we have $\rho(y, f(y)) < \mu(\epsilon)$.

We will need the following standard formula from spherical trigonometry. Consider a geodesic triangle with sides A, B, C and opposite angles α, β, γ on the sphere $S^n_{1/\sqrt{\delta}}$ of curvature δ. Then

$$\cos A\sqrt{\delta} = \cos B\sqrt{\delta} \cdot \cos C\sqrt{\delta} + \sin B\sqrt{\delta} \cdot \sin C\sqrt{\delta} \cdot \cos \alpha. \tag{7.6}$$

Proof of Lemma 7.5. It suffices to consider the case y not antipodal to y_1. Let y_1 be the mid-point of the geodesic segment which goes from y_0 to its antipodal point and passes through y. Consider the triangle with vertices $y_0, f(y), y$ and the triangle with vertices $y_0, f(y), y_1$. Denote by θ the angle at y_0 which is common to these triangles. (See Fig. 7.3).

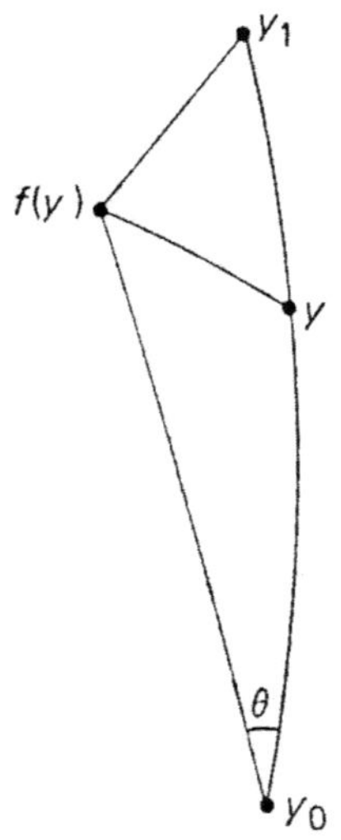

Fig. 7.3.

Applying (7.6) gives

$$\begin{aligned} \cos \rho(y, f(y)) = {} & \cos \rho(y_0, y) \cos \rho(y_0, f(y)) \\ & + \sin \rho(y_0, y) \sin \rho(y_0, f(y)) \cos \theta. \end{aligned} \tag{7.7}$$

Also, using the formula again gives

$$\cos \rho(y_1, f(y)) = \sin \rho(y_0, f(y)) \cos \theta \tag{7.8}$$

because $\rho(y_0, y_1) = \pi/2$.

Substituting (7.8) into (7.7) gives

$$\cos \rho(y, f(y)) = \cos \rho(y_0, y) \cos \rho(y_0, f(y)) + \sin \rho(y_0, y) \cos \rho(y_1, f(y)). \tag{7.9}$$

Now by the triangle inequality,

$$|\rho(y_0, f(y)) - \rho(f(y_0), f(y))| < \rho(y_0, f(y_0)) \tag{7.10}$$

$$|\rho(y_1, f(y)) - \rho(f(y), f(y_1))| < \rho(y_1, f(y_1)). \tag{7.11}$$

Using (1) and (2) of the hypothesis, (7.10) and (7.11) imply

$$|\rho(y_0, f(y)) - (1 \pm \epsilon)\rho(y_0, y)| < \epsilon \tag{7.12}$$

$$|\rho(y_1, f(y)) - (1 \pm \epsilon)\rho(y, y_1)| < \epsilon. \tag{7.13}$$

Substituting (7.12), (7.13) into (7.9) and using $\frac{1}{2}\pi - \rho(y_0, y) = \pm\rho(y, y_1)$ gives

$$\begin{aligned} \cos\rho(y, f(y)) &\approx \cos\rho(y_0, y)\cos[(1 \pm \epsilon)\rho(y_0, y) + \epsilon] \\ &\quad + \sin\rho(y_0, y)\cos[\pm((1 \pm \epsilon)(\frac{1}{2}\pi - \rho(y_0, (y))) + \epsilon]. \end{aligned} \tag{7.14}$$

Using

$$\cos(\frac{1}{2}\pi - \rho(y_0, y)) = \sin\rho(y_0, y), \text{ and } \cos^2\rho(y_0, y) + \sin^2\rho(y_0, y) = 1,$$

this clearly suffices to complete the proof. □

Up to this point, our preliminaries have been purely topological. To apply these results, we will need the following estimate, which compares the behavior of a Jacobi field on a δ-pinched manifold with that of a Jacobi field on the unit sphere. We introduce the notation $K(\delta)$ for any positive decreasing function defined on (0,1) whose limit as $\delta \to 1$ is 0.

LEMMA 7.15. *Suppose* $1 \geq K_M > \delta > 0$. *Let* $\gamma : [0, t] \to M$ *be a normal geodesic and* J *be a Jacobi field along* γ *with* $J(0) = 0$ *and* $\langle J, \gamma' \rangle \equiv 0$. *Set*

$$P_{\gamma(t)}(J'(0)) = E(t).$$

Then for $0 < t < \pi$,

$$||J(t) - \sin tE(t)|| \leq (1 - \delta)(e^t - 1 - t)||J'(0)|| = K(\delta).$$

PROOF. Set $\sin t\, E(t) = Z(t)$ and $J - Z = W$. The Jacobi field equation for J may be written in matrix form as a first order system

$$\begin{bmatrix} J \\ J' \end{bmatrix}' = \begin{bmatrix} 0 & I \\ -R_t & 0 \end{bmatrix} \begin{bmatrix} J \\ J' \end{bmatrix},$$

where $R_t(v) = R(v, \gamma'(t))\gamma'(t)$. Also, Z satisfies

$$\begin{bmatrix} Z \\ Z' \end{bmatrix}' = \begin{bmatrix} 0 & I \\ -I & 0 \end{bmatrix} \begin{bmatrix} Z \\ Z' \end{bmatrix}.$$

Then

$$\begin{aligned} \begin{bmatrix} W \\ W \end{bmatrix}' &= \begin{bmatrix} 0 & -I \\ I & 0 \end{bmatrix} \begin{bmatrix} W \\ W' \end{bmatrix} + \begin{bmatrix} 0 & 0 \\ I - R_t & 0 \end{bmatrix} \begin{bmatrix} J \\ J' \end{bmatrix} \\ &= \begin{bmatrix} 0 & -I \\ I & 0 \end{bmatrix} \begin{bmatrix} W \\ W' \end{bmatrix} + \begin{bmatrix} 0 \\ (I - R_t)(J) \end{bmatrix}. \end{aligned}$$

Notice that for x of unit length,

$$1 - \langle R_t(x), x\rangle = 1 - \langle R(x,\gamma')\gamma', x\rangle \leq (1-\delta).$$

Since $x \to R(x,\gamma')\gamma'$ is self-adjoint, it follows that

$$||(I - R_t)|| \leq 1 - \delta.$$

Then by the above

$$\left\| \begin{bmatrix} W \\ W' \end{bmatrix} \right\|' \leq \left\| \begin{bmatrix} W \\ W' \end{bmatrix}' \right\| \leq \left\| \begin{bmatrix} W \\ W' \end{bmatrix} \right\| + (1-\delta)||J'(0)||t,$$

since by Rauch I (comparing with $\mathbb{R}^n$),

$$||J(t)|| \leq ||J'(0)||t.$$

Since $W(0) = W'(0) = 0$, integrating the above yields

$$\left\| \begin{bmatrix} W \\ W' \end{bmatrix} \right\| \leq (1-\delta)\frac{1}{2}t^2||J'(0)|| + \int_0^t \left\| \begin{bmatrix} W \\ W' \end{bmatrix} \right\| \mathrm{d}s.$$

Iterating this estimate yields

$$\left\| \begin{bmatrix} W \\ W' \end{bmatrix} \right\| \leq (1-\delta)\Big(\frac{t^2}{2} + \cdots + \frac{t^n}{n!}\Big)||J'(0)|| + \int_0^t \int_0^s \cdots \left\| \begin{bmatrix} W \\ W' \end{bmatrix} \right\| \ldots \mathrm{d}s.$$

Let

$$M = \max_{[0,t]} \left\| \begin{bmatrix} W \\ W' \end{bmatrix} \right\|.$$

Then the second term on the right is smaller than $Mt^{n-1}/(n-1)!$. Thus letting $n \to \infty$,

$$M = \max_{[0,t]} \left\| \begin{bmatrix} W \\ W' \end{bmatrix} \right\| \leq (1-\delta)(e^t - 1 - t)||J'(0)||.$$

□

We are now ready to state our main result.

THEOREM 7.16. *There exists $1 > \delta \geq \frac{1}{4}$ such that if M^n is a complete simply connected Riemannian manifold satisfying $1 \geq K_M > \delta$, then M^n is diffeomorphic to the n-sphere S^n.*

In order to prove Theorem 7.16 we will need three lemmas. We begin by recalling the proof of the Sphere Theorem. In Chapter 6 it was shown that for $p, q \in M^n$ at maximal distance and any normal geodesic τ emanating from p there exists a unique value t_0 such that

$$\frac{1}{2}\pi \leq \rho(\tau(t_0), p) = \rho(\tau(t_0), q) \leq \frac{\pi}{2\sqrt{\delta}}.$$

Let E be the set of such $\tau(t_0)$, which is precisely the set of points which are equidistant from p and q. Set

$$\rho_p(x) = \rho(x,p), \qquad \rho_q(x) = \rho(x,q).$$

Since $i(M) \geq \pi$, ρ_p and ρ_q are smooth in a neighborhood of $E = (\rho_p - \rho_q)^{-1}(0)$. Since $\operatorname{grad}\rho_p = \tau'(t_0)$ and $\operatorname{grad}\rho_q = \tau_0'(t_0)$, where τ_0 is the unique normal minimal geodesic from q to $\tau(t_0)$, it follows that 0 is the regular value of $\rho_p - \rho_q$ and E is a smooth submanifold. It follows easily from the first variation formula that the tangent space $E_{\tau(t_0)}$ of E consists of the hyperplane of vectors making equal angles with $-\tau'(t_0)$ and $-\tau_0'(t_0)$. Let $\Phi_p(x) = \exp_p^{-1} x/\|\exp_p^{-1} x\|$. Then Φ_p is smooth on $B_\pi(p) - p$ and $\Phi_p|E$ is given by $\tau(t_0) \to \tau'(0)$. Define Φ_q similarly on $B_\pi(q) - q$. Then if $w \in E_{\tau(t_0)}$, and

$$Z(t) = J(t) + atT$$

is the unique Jacobi field along τ such that $Z(0) = 0$ and $Z(t_0) = w$, we find that $d\Phi_p(w) = J'(0)$. In particular $d\Phi_p, d\Phi_q$ are nonsingular. Hence Φ_p^{-1}, Φ_q^{-1} are diffeomorphisms from S^{n-1} to E. Then

$$t_0 = \rho_p \circ \Phi_p^{-1} = \rho_q \circ \Phi_q^{-1}$$

is a smooth function of $\tau'(0)$ and of $\tau_0'(0)$. We now have easily:

LEMMA 7.17. *Let M^n be as in Theorem 7.16. Then M^n is a twisted sphere.*

PROOF. Let

$$\eta(t,s) : [0,1] \times \left[\frac{\pi}{2}, \frac{\pi}{2\sqrt{\delta}}\right] \to \mathbb{R}$$

be a smooth function satisfying $\partial\eta/\partial t > 0$ and

$$\eta(t,s) = \begin{cases} t, & 0 \leq t \leq \frac{1}{4}, \\ ts, & \frac{3}{4} \leq t \leq 1. \end{cases}$$

Choose isometries

$$I_1 : (\mathbb{R}^n)_0 \to M_p^n, \qquad I_2 : (\mathbb{R}^n)_0 \to M_q^n.$$

Identify $\mathbb{R}^n$ with $(\mathbb{R}^n)_0$ and set

$$\psi_p(x) = \rho_p \circ \Phi_p^{-1}(I_1(x/\|x\|)), \qquad \psi_q(x) = \rho_q \circ \Phi_q^{-1}(I_2(x/\|x\|)).$$

Define

$$h_1(x) = \exp_p \eta(\|x\|, \psi_p(x)) I_1(x), \qquad h_2(x) = \exp_q \eta(\|x\|, \psi_q(x)) I_2(x).$$

Then for sufficiently small ϵ it follows that $h_1, h_2|B_{1+\epsilon}^n$ satisfy the conditions of Definition 7.1. □

Fix a minimal geodesic γ from p to q and I as above. Let $s_{\gamma'}$ denote reflection in the hyperplane perpendicular to $\gamma'(d)$, where $d = \rho(p,q)$. From now on, we will always choose $I_2 = s_{\gamma'} \circ P_\gamma \circ I_1$. Note that if M^n is actually

S^n and $\mathrm{d}h_i = I_i$, then this choice coincides with the standard covering of S^n used in Proposition 7.2.

Let $g = \Phi_q \circ \Phi_p^{-1}$ denote the map defined by $\tau'(0) \to \tau_0'(0)$. Then since $\eta(1, s) = s$, we see that

$$f(x) = h_2^{-1} \circ h_1 | S^{n-1} = I_2^{-1}\Big(\exp_q^{-1} \Big(\frac{\exp_p \psi_p(x) I_1(x)}{||\psi_q(x)||} \Big)\Big)$$
$$= I_2^{-1} \circ g \circ I_1 = I_1^{-1} \circ P_\gamma^{-1} \circ s_{\gamma'}^{-1} \circ g \circ I_1.$$

Since I_1, P_γ and $s_{\gamma'}$ are isometries, we have

$$||f - 1|| = ||s_{\gamma'} \circ P_\gamma - g||$$

where 1 means the identity map. Also for all y,

$$||\mathrm{d}f_y - 1|| = ||s_{\gamma'} \circ P_\gamma - \mathrm{d}g_y||.$$

Therefore, to estimate $||f - 1||$ and $||\mathrm{d}f_y - 1||$, it suffices to estimate

$$||s_{\gamma'} \circ P_\gamma - g|| \text{ and } ||s_{\gamma'} \circ P_\gamma - \mathrm{d}g_y||.$$

In fact, we will estimate

$$||s_{\tau_0'} \circ P_{\tau \cup \tau_0} - g||, \text{ and } ||s_{\tau'} \circ P_{\tau \cup \tau_0} - \mathrm{d}g_y||$$

for any τ.

LEMMA 7.18. *Let M^n be as in Theorem 7.16 and let $\tau'(0) \in S_p^{n-1}$. Then*

$$\mathrm{d}g_{\tau'(0)} : (S^{n-1})_{\tau'(0)} \to (S_q^{n-1})_{\tau_0'(0)}$$

satisfies

$$||s_{\tau_0'} \circ P_{\tau \cup \tau_0} - \mathrm{d}g_{\tau'(0)}|| \le K(\delta).$$

PROOF. Consider the geodesic triangle with vertices at p, q and $\tau(t_0) = \tau_0(t_0)$. Then $\pi \le \rho(p, q) \le \pi/\sqrt{\delta}$ and $\frac{\pi}{2} \le t_0 \le \frac{\pi}{2\sqrt{\delta}}$. We begin by estimating $\sphericalangle(-\tau'(t_0), \tau_0'(t_0)) = \theta$ by use of Toponogov's Theorem. The comparison triangle on $S^{n-1}_{1/\sqrt{\delta}}$ for which $\bar{\theta}$ is maximal, has side lengths $\frac{\pi}{2\sqrt{\delta}}, \frac{\pi}{2\sqrt{\delta}}, \pi$. Applying (7.6) gives the estimate

$$\pi\sqrt{\delta} \le \theta \le \pi. \tag{7.19}$$

Let $v \in (S_p^{n-1})_{\tau'(0)}, ||v|| = 1$, and let $Z(t) = J(t) + atT$ be the Jacobi field such that $J(0) = 0, J'(0) = v$ and $Z(t_0) \in E_{\tau(t_0)}$. Let $Z_0(t)$ be the Jacobi field along τ_0 such that

$$Z_0(0) = 0, \qquad Z_0(t_0) = Z(t_0).$$

Then since $E_{\tau(t_0)}$ makes equal angles with $-\tau'(t_0)$ and $-\tau_0'(t_0)$, it follows that

$$Z_0(t) = J_0(t) + atT_0(t), \qquad ||J(t_0)|| = ||J_0(t_0)||.$$

It follows from the remarks preceding Lemma 7.17 that

$$\mathrm{d}g_{\tau'(0)}(v) = J_0'(0).$$

Moreover, (7.19) implies

$$|a| \leq \cos \frac{1}{2}\pi\sqrt{\delta} = K(\delta). \tag{7.20}$$

Now by Rauch I, we have the estimate

$$||J_0(t_0)|| = ||J(t_0)|| \leq \sin(\delta t_0).$$

Similarly,

$$(\sin t_0)||J_0'(0)|| \leq ||J_0(t_0)||.$$

Combining these gives

$$||J_0'(0)|| \leq \frac{\sin(\delta t_0)}{\sin t_0} \leq 1 + K(\delta). \tag{7.21}$$

By Lemma 7.15 and (7.21) we have

$$|| \sin t_0 E(t_0) - J(t_0)|| \leq K(\delta)$$

$$||J_0(t_0) - \sin t_0 E_0(t_0)|| \leq K(\delta)(1 + K(\delta)) = K(\delta), \tag{7.22}$$

where

$$E(t) = P_{\tau(t)}(J'(0)), \qquad E_0(t) = P_{\tau_0(t)}(J'(0)).$$

Since $\mathrm{d}g_{\tau'(0)}(v) = J_0'(0)$, we have

$$||P_{\tau\cup-\tau_0}(v) - \mathrm{d}g_{\tau'(0)}(v)|| = ||E(t_0) - E_0(t_0)||. \tag{7.23}$$

But since $Z(t_0) = Z_0(t_0)$, we have

$$\begin{aligned} |\sin t_0|||E(t_0) - E_0(t_0)|| &\leq || \sin t_0 E(t_0) - Z(t_0)|| + ||Z(t_0) - \sin t_0 E_0(t_0)|| \\ &\leq || \sin t_0 E(t_0) - J(t_0)|| + ||J_0(t_0) - \sin t_0 E_0(t_0)|| \\ &\quad + |a||t_0|||T_0(t_0) - T(t_0)||. \end{aligned}$$

Using (7.22) to estimate the first two terms on the right-hand side and (7.20) to estimate the third gives

$$|\sin t_0|||E(t_0) - E_0(t_0)|| \leq K(\delta).$$

This estimate together with (7.23) and $\frac{\pi}{2} \leq t_0 \leq \frac{\pi}{2\sqrt{\delta}}$ completes the proof. ☐

LEMMA 7.24. *Let M^n be as in Theorem 7.16 and let $\tau'(0) \in S_p^{n-1}$. Then, for $u = \tau'(0)$ or $\langle u, \tau'(0)\rangle = 0$, we have*

$$||s_{\tau_0'} \circ P_{\tau\cup-\tau_0}(u) - g(u)|| \leq K(\delta).$$

PROOF. If $u = \tau'(0)$, then the estimate is obvious from Lemma 7.18. So suppose $\langle u, \tau'(0)\rangle = 0$. Let $E(t) = P_{\tau(t)}(u)$, and define s_0 by

$$\exp_p(s_0 u) = \exp_q(s_0 g(u)).$$

Set

$$c(t) = \exp_{\tau(t)}(s_0 E(t)).$$

We use Corollary 1.36 of Rauch II to compare the length of c to that of the corresponding curve (which is a circle of latitude) on the sphere $S^n_{1/\sqrt{\delta}}$. This gives the estimate

$$L[c] \leq \frac{t_0}{2\pi/\sqrt{\delta}} \frac{1}{\sqrt{\delta}} \sin\Big(\sqrt{\delta}(\frac{1}{2}\pi\sqrt{\delta} - s_0)\Big).$$

Using $\frac{\pi}{2} \leq s_0 \leq \frac{\pi}{2\sqrt{\delta}}$, we find

$$L[c] \leq K(\delta). \tag{7.25}$$

Let $E_0(t)$ be the parallel field along τ_0 such that $E_0(t_0)$ is the vector perpendicular to $\tau_0'(t_0)$ which makes the smallest angle with $E(t_0)$. Then if

$$c_0(t) = \exp_{\tau_0(t)}(s_0 E_0(t))$$

as above, we have

$$L[c_0] \leq K(\delta). \tag{7.26}$$

Now $P_{-\tau_0}(E(t_0)) = P_{\tau\cup-\tau_0}(u)$. The inequality of Lemma 7.18 implies

$$\sphericalangle(E(t_0), E_0(t_0)) = \sphericalangle(P_{\tau\cup-\tau_0}(u), E_0(0)) = K(\delta).$$

Then Toponogov's Theorem gives the estimate

$$\begin{aligned} \rho(\exp_{\tau(t_0)} s_0 E(t_0), \exp_{\tau(t_0)} s_0 E_0(t_0)) &\leq (1/\sqrt{\delta})K(\delta) = K(\delta), \\ \rho(\exp_{\tau_0(t_0)} s_0 P_{\tau\cup-\tau_0}(u), \exp_{\tau_0(0)} s_0 E_0(0)) &\leq (1/\sqrt{\delta})K(\delta) = K(\delta). \end{aligned} \tag{7.27}$$

Adding (7.25), (7.26), (7.27) gives

$$\rho(\exp_p s_0 u, \exp_q s_0 P_{\tau\cup-\tau_0}(u)) \leq K(\delta).$$

Therefore

$$\rho(\exp_q s_0 g(u), \exp_q s_0 P_{\tau\cup-\tau_0}(u)) \leq K(\delta). \tag{7.28}$$

Now choose $K(\delta)$ in (7.28) to be smaller than $2\pi\Big(1 - \frac{1}{2\sqrt{\delta}}\Big)$. Then the minimal geodesic σ from $\exp_q(s_0 g(u))$ to $\exp_q(s_0 P_{\tau\cup-\tau_0}(u))$ is contained in $B_\pi(q)$. Set $\alpha = \sphericalangle(g(u), P_{\tau\cup-\tau_0}(u))$. Since $i(M^n) \geq \pi$, we may apply Corollary 1.35 of Rauch I (comparing with S^n_1) and (7.6) to get

$$K(\delta) \geq \cos^{-1}\Big(\cos^2\frac{\pi}{2\sqrt{\delta}} + \sin^2\frac{\pi}{2\sqrt{\delta}}\cos\alpha\Big).$$

Therefore

$$\alpha \leq \cos^{-1}\Big(\frac{\cos K(\delta) - \cos^2\frac{\pi}{2\sqrt{\delta}}}{\sin^2\frac{\pi}{2\sqrt{\delta}}}\Big) \leq K(\delta). \tag{7.29}$$

This suffices to complete the proof. □

The proof of Theorem 7.16 is now easy.

PROOF OF THEOREM 7.16. By Lemma 7.17, M^n is a twisted sphere. By Lemma 7.5, Lemma 7.18 and Lemma 7.24 we have for any τ,

$$||g - s_{\tau'} \circ P_{\tau \cup - \tau_0}|| \leq K(\delta), \tag{7.30}$$

$$||s_{\tau'} \circ P_{\tau \cup - \tau_0} - \mathrm{d}g_{\tau'(0)}|| \leq K(\delta). \tag{7.31}$$

Thus if γ is the minimal geodesic from p to q, (7.31) says in particular that

$$||s_{\gamma'} \circ P_\gamma - g|| \leq K(\delta). \tag{7.32}$$

Restricting to $S^{n-1}_{\tau'(0)}$ and adding (7.30), (7.31), (7.32) gives

$$||s_{\gamma'} \circ P_\gamma - \mathrm{d}g_{\tau'(0)}|| \leq K(\delta). \tag{7.33}$$

As we have observed before Lemma 7.18, (7.32) and (7.33) imply that for all y,

$$||f - 1|| \leq K(\delta), \qquad ||\mathrm{d}f_y - 1|| \leq K(\delta).$$

For δ sufficiently close to 1, $K(\delta)$ will be so small that f satisfies the hypothesis of Proposition 7.4. Then f is isotopic to the identity and by Proposition 7.2, M^n is diffeomorphic to S^n. □

We will now discuss without proof some results which may serve to put the theorems of Chapter 6 and the present chapter in perspective. As the theorems suggest, what we would really like to do is to classify compact manifolds of positive curvature. This problem is almost entirely open. Corollary 1.32, Theorem 5.9 and Theorem 8.23 (for nonnegative curvature) provide some information about the stucture of the fundamental group. However, it was only recently that Singer and Lichnerowicz, by use of the Index Theorem, gave the first result in the simply connected case. They showed that the $\hat{A}$ genus of a compact manifold which admits a metric of positive curvature vanishes. In fact, this holds if only the scalar curvature is positive. Their theorem provides the first examples of compact simply connected manifolds which can be shown not to admit a metric of positive curvature. On the other hand, Example 3.44 describes the best attempt so far to construct a simply connected nonhomogeneous manifold which actually does admit a metric of positive curvature. Moreover there are only a few conjectures concerning the properties of such manifolds. One such is the Hopf conjecture, that the product of compact manifolds does not admit a metric of positive curvature. A related conjecture of Hopf and Chern says that the Euler characteristics of an even-dimensional manifold admitting a metric of positive (nonnegative) curvature is positive (nonnegative). This is known in dimensions 2 and 4.

In view of the lack of information in the general case, it is natural to see what can be said about manifolds admitting a δ-pinched metric, for various specific values of δ. The theorems of Chapter 6 and the present chapter essentially provide a classification for $\delta = \frac{1}{4}$. For smaller values of δ there are fragmentary results which are proved by analytic (Hodge theory) techniques. In even dimensions at least there is the following result.

THEOREM 7.34 (Cheeger [1970], Weinstein [1967]). *Given* n *and* $\delta > 0$, *there are at most finitely many diffeomorphism classes of* $2n$*-manifolds admitting a metric such that* $1 \geq K_M \geq \delta$.

In odd dimensions Theorem 7.34 is false without assuming simple connectivity (lens spaces), and even in that case it is not known. The reason for this is that the proof depends on Corollary 1.32 and Theorem 5.9 (which holds only in even dimensions).

The problem of studying a δ-pinched manifold may be thought of as that of studying a manifold whose curvature is similar to that of the sphere. Then a natural generalization is to replace the sphere by an arbitrary compact simply connected Riemannian manifold. More explicitly, we would like to weaken the hypothesis of the Cartan-Ambrose-Hicks Theorem 1.42 to some sort of topological equivalence. In order to do this, the behavior of ∇R, the covariant derivative of the curvature tensor, must be taken into account. We make the following definition.

DEFINITION 7.35. Let M, M_0 be as in Theorem 1.42. Given $N > 0$, define

$$\rho_N(M, M_0) = \sup_{\gamma}\{||R - I_\gamma^{-1}(R_0)||, ||\nabla R - I_\gamma^{-1}(\nabla R_0)||\},$$

where the sup is taken over all broken geodesics with at most N breaks, each segment of which has length ≤ 1.

As usual, let $V(M)$ denote the volume of M.

THEOREM 7.36 (Cheeger [1967]). *Given a compact simply connected Riemannian manifold* M^n, *there exists* $N > 0$ *such that given* $V > 0$ *there exists* $\epsilon > 0$ *such that if* M_0 *is a compact simply connected Riemannian manifod for which* $V(M_0) > V$ *and* $\rho_N(M, M_0) < \epsilon$, *then* M_0 *is diffeomorphic to* M.

The proof of Theorem 7.36 proceeds by a monodromy argument similar to the proof of the Cartan-Ambrose-Hicks Theorem. As we have seen, in certain special cases such as Theorem 7.16, the hypothesis on ∇R, ∇R_0 can be removed. In fact, it can be shown that this can be done when M^n is any symmetric space of rank one. However, at present, one cannot get any topological conclusion at all in the general case without such a hypothesis. But there is the following analogue of Theorem 7.34.

THEOREM 7.37 (Cheeger [1970]). *Given* $n, d, V, K > 0$, *there are at most finitely many diffeomorphism classes of* n*-manifolds which admit a metric such that* $d(M) < d$, $V(M) > V$, $|K_M| < K$.

CHAPTER 8

Complete Manifolds of Nonnegative Curvature

In this chapter we shall investigate the topological and geometrical properties of complete noncompact manifolds of nonnegative curvature. For further discussion see Cheeger and Gromoll [1968, 1971, 1972], Gromoll and Meyer [1969] and Milnor [1968]. We will show first of all that a manifold M with $K_M \geq 0$ contains a compact totally geodesic submanifold S. Actually, S has a much stronger property of being totally convex in the sense of Definition 8.1. S is called the *soul* of M. The existence of a totally geodesic submanifold is remarkable in view of the fact that most Riemannian manifolds do not contain nontrivial totally geodesic submanifolds. Furthermore, we shall see that the inclusion $i : S \to M$ is a homotopy equivalence. More generally this is true for any totally convex set $C \subset M$. Thus in particular the noncompact manifold M has the homotopy type of a compact manifold. If $K_M > 0$, S reduces to a point. With more technical work (see Cheeger and Gromoll [1972]) one can show that M is actually diffeomorphic to the normal bundle $\nu(S)$ of S. So $K_M > 0$ implies M is diffeomorphic to $\mathbb{R}^n$. The diffeomorphism does not in general arise from the exponential map.

DEFINITION 8.1. A set C is called *totally convex* if whenever $p, q \in C$ and γ is a geodesic segment from p to q, then $\gamma \subset C$.

It is clear from the point of view of Morse Theory that totally convex sets are of interest, since all critical points of the energy function on $\Omega_{p,q}(M)$ are already contained in C. Therefore C should be topologically similar to M in some sense. Proper totally convex sets ($C \neq M$) do not exist in most manifolds. In Euclidean space, any convex set is an intersection of half-spaces. We now give a way of constructing a half-space which makes sense in any noncompact Riemannian manifold.

A *ray* $\gamma : [0, \infty) \to M$ is a geodesic parmeterized by arclength each finite segment of which realizes the distance between its endpoints.

In any noncompact manifold M there is at least one ray starting at each point of M. Given p, let p_i be a sequence of points such that $\rho(p, p_i) \to \infty$. Let γ_i be a sequence of minimal geodesics from p to p_i, and let v be an accumulation point of the vectors $\gamma_i'(0)$. Let $\gamma : [0, \infty) \to M$ be the geodesic such that $\gamma'(0) = v$. Then for any a, $\gamma|[0, a]$ is minimal. In fact $\gamma|[0, a]$ is the uniform limit of the segments $\gamma_i|[0, a]$, which are all minimal for sufficiently large i. Thus $\gamma|[0, a]$ is minimal so γ is a ray. Let $B_r(p)$ as usual denote the open metric ball of radius r centered at p. Given γ, we define B_γ to be $\bigcup_t B_t(\gamma(t))$ and H_γ to be the complement $(B_\gamma)'$ of B_γ. In Euclidean space, H_γ is a half-space in the usual sense. Notice also that since the balls $B_t(\gamma(t))$ are open, $\gamma(0) \in (B_\gamma)' = H_\gamma$. Hence H_γ is nonempty.

THEOREM 8.2. *Let M be a noncompact manifold of nonnegative curvature. Then for any ray γ, H_γ is totally convex.*

The reader should compare our construction to those in Lemma 6.10. In both situations we are making use of the principle that positive curvature in M tends to make the complement of a sufficiently large metric ball convex.

PROOF. We use Toponogov's Theorem. Suppose there exists a geodesic $\gamma_0 : [0,1] \to M$ with endpoints $\gamma_0(0), \gamma_0(1) \in (B_\gamma)'$, but $\gamma_0(s) \in B_\gamma$ for some $s \in (0,1)$. It follows from the triangle inequality that $t_2 \geq t_1 > 0$ implies $B_{t_2}(\gamma(t_2)) \supset B_{t_1}(\gamma(t_1))$, and hence that for $q = \gamma_0(s) \in B_\gamma$ there exists $t_0 > 0$ such that $q \in B_t(\gamma(t)), t \geq t_0$. In fact setting

$$t_0 - \epsilon = \rho(q, \gamma(t_0)), \qquad \epsilon > 0,$$

we have

$$\begin{aligned}\rho(q,\gamma(t)) &\leq \rho(q,\gamma(t_0)) + \rho(\gamma(t_0),\gamma(t)) \\ &= t_0 - \epsilon + t - t_0 = t - \epsilon\end{aligned}$$

for all $t \geq t_0$.

Let $\gamma_0(s_t)$ be a point on γ_0 which is closest to $\gamma(t)$. Further, consider the restriction $\gamma_0^t = \gamma_0|[0, s_t]$, and minimal geodesics γ_1^t from $\gamma_0(s_t)$ to $\gamma(t)$, γ_2^t from $\gamma(t)$ to $\gamma_0(0)$. It follows from the above that for all $t \geq t_0$,

$$L[\gamma_2^t] \geq L[\gamma_1^t] + \epsilon.$$

On the other hand, since $L[\gamma_0^t] < L[\gamma_0]$ is finite, we will have

$$L[\gamma_2^t] + L[\gamma_1^t] > L[\gamma_0^t],$$

for t sufficiently large. Hence we may construct a triangle in Euclidean 2-space as in Toponogov's Theorem(A). In particular, for this triangle we have $\bar{\alpha}_2^t \leq \alpha_2^t$.

Using the law of cosines in the Euclidean plane, we obtain

$$\begin{aligned}\cos \bar{\alpha}_2^t &= \frac{L^2[\gamma_0^t] + L^2[\gamma_1^t] - L^2[\gamma_2^t]}{2L[\gamma_0^t]L[\gamma_1^t]} \\ &= \frac{L[\gamma_1^t] + L[\gamma_2^t]}{2L[\gamma_1^t]} \frac{L[\gamma_1^t] - L[\gamma_2^t]}{2L[\gamma_0^t]} + \frac{L[\gamma_0^t]}{2L[\gamma_1^t]}.\end{aligned}$$

Now

$$L[\gamma_2^t] - L[\gamma_1^t] \geq \epsilon, \qquad L[\gamma_0^t] < L[\gamma_0], \qquad L[\gamma_2^t] \geq t,$$

so for sufficiently large t we have $\cos \bar{\alpha}_2^t < 0$, implying $\frac{1}{2}\pi < \bar{\alpha}_2^t \leq \alpha_2^t$. However, $\gamma_0(s_t)$ is closest to $\gamma(t)$ and $s_t \in (0,1)$, so $\alpha_2^t = \frac{1}{2}\pi$, contradiction. □

EXAMPLE 8.3. Consider the paraboloid of revolution $z = x^2 + y^2$ (see Fig. 8.1). This is a manifold whose curvature is strictly positive (but not bounded away from zero). One can check that the rays on the paraboloid are precisely the meridians $z = (1 + \alpha^2)x^2$.

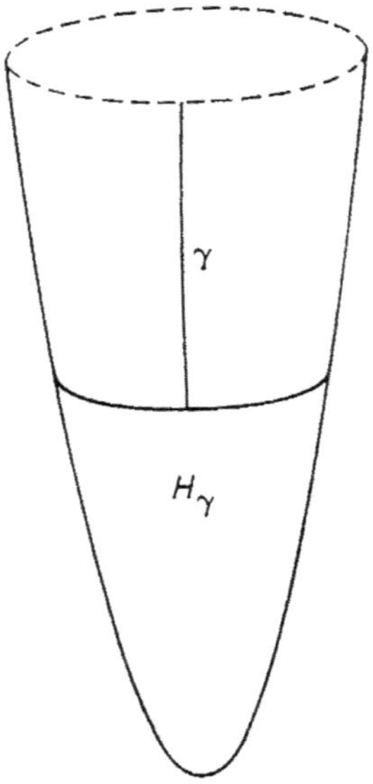

Fig. 8.1.

The half-spaces H_γ are the caps indicated in Fig. 8.1. We leave the verification as a (nontrivial) exercise.

EXAMPLE 8.4. Consider the cylinder $x^2 + y^2 = 1$ (see Fig. 8.2). The rays, as we have seen in Example 2.4, are the geodesics parallel to the z-axis. The half-spaces are the pieces of the cylinder below or above some plane $z =$ constant.

Once we have constructed a single nontrivial totally convex set (t.c.s.) it is easy to construct a compact one. What we want to look at eventually are t.c.s. which are in some sense as small as possible.

PROPOSITION 8.5. *With M as above and $p \in M$, there exists a family of compact totally convex sets $C_t, t \geq 0$, such that*

(1) $t_2 \geq t_1$ implies $C_{t_2} \supset C_{t_1}$ and

$$C_{t_1} = \{q \in C_{t_2} \mid \rho(q, \partial C_{t_2}) \geq t_2 - t_1\}.$$

In particular

$$\partial C_{t_1} = \{q \in C_{t_2} \mid \rho(q, \partial C_{t_2}) = t_2 - t_1\}.$$

(2) $\bigcup_{t \geq 0} C_t = M$,

(3) $p \in \partial C_0$.

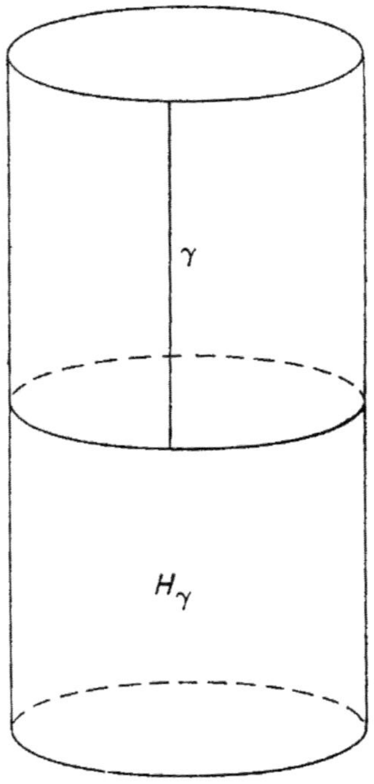

Fig. 8.2.

Here our notation is $\partial A = \mathrm{cl}A \cap \mathrm{cl}A' = \mathrm{cl}A - \mathrm{int}A$, where $\mathrm{cl}A$ denotes the closure and $\mathrm{int}A$ the interior of a subset A in M.

PROOF. Set $C_t = \bigcap_\gamma (B_{\gamma_t})'$, where the intersection is taken over all rays γ emanating from p, and $\gamma_t : [0, \infty) \to M$ denotes the restricted ray from $\gamma(t)$ to ∞ with $\gamma_t(s) = \gamma(s+t)$. Clearly C_t is totally convex and closed. If some C_t were not compact, it would contain a sequence of points $p_i \to \infty$. Let $\gamma_i : [0, \beta_i] \to M$ be a sequence of minimal normal geodesics from p to p_i. Each γ_i is contained in the t.c.s. C_t. Then by compactness of the unit sphere in M_p, we may choose a subsequence of γ_i converging to a ray $\gamma : [0, \infty) \to C_t$. But then by definition, $\gamma(t') \notin C_t$ for $t' > t$. Hence for large i, $p_i \notin C_t$. which is a contradiction. It follows that C_t is compact.

(1) Assume that $t_2 > t_2$. Since

$$B_s(\gamma_{t_2}(s)) = B_s(\gamma(s+t_2)) \subset B_{s+t_2-t_1}(\gamma(s+t_2)) = B_{s+t_2-t_1}(\gamma_{t_1}(s+t_2-t_1)),$$

we obtain that $B_{\gamma_{t_2}} \subset B_{\gamma_{t_1}}$ for any ray γ and hence $C_{t_2} \supset C_{t_1}$.

Given a ray γ from p, we have

$$B_{\gamma_{t_1}} = \{q \mid \rho(q, B_{\gamma_{t_2}}) < t_2 - t_1\}.$$

For if $\rho(q, B_{\gamma_{t_2}}) < t_2 - t_1$, then there is $q' \in B_{\gamma_{t_2}}$ such that $\rho(q, q') < t_2 - t_1$. So when $s > 0$ is sufficiently large, $q' \in B_s(\gamma_{t_2}(s))$ and, by the triangle inequality,

$$q \in B_{s+t_2-t_1}(\gamma(s+t_2)) \subset B_{\gamma_{t_1}}.$$

Conversely, if $q \in B_{\gamma_{t_1}}$, then, for some $s > 0$,

$$q \in B_{s+t_2-t_1}(\gamma_{t_1}(s+t_2-t_1)) = B_{s+t_2-t_1}(\gamma_{t_2}(s)) \supset B_s(\gamma_{t_2}(s)).$$

Hence

$$\rho(q, B_{\gamma_{t_2}}) \leq \rho(q, B_s(\gamma_{t_2}(s))) < t_2 - t_1.$$

If A_i is an arbitrary collection of nonempty subsets of M, then

$$\rho\Big(q, \bigcup A_i\Big) < t_2 - t_1$$

if and only if there exists j with $\rho(q, A_j) < t_2 - t_1$. Therefore

$$\bigcup_\gamma B_{\gamma_{t_1}} = \Big\{q \,|\, \rho\Big(q, \bigcup_\gamma B_{\gamma_{t_2}}\Big) < t_2 - t_1\Big\},$$

where γ ranges over all rays emanating from p, and hence

$$C_{t_1} = \Big\{q \,|\, \rho\Big(q, \bigcup_\gamma B_{\gamma_{t_2}}\Big) \geq t_2 - t_1\Big\}.$$

Note that for any $A \subset M, q \in A'$ one has $\rho(q, A) = \rho(q, \partial A)$, and $\partial A = \partial A'$. Let $A = \bigcup_\gamma B_{\gamma_{t_2}}$. This completes the proof.

(2) It follows from the triangle inequality that for any $q \in M$ we have $q \in C_t$ whenever $t \geq \rho(p, q)$.

(3) This is an immediate consequence of the construction. □

We will now show that for any closed totally convex set C, there exists a set $N \subset C$ which has the structure of a smooth connected embedded totally geodesic submanifold such that $C \subset \bar{N}$. This result may be thought of as a sort of regularity theorem and is necessary to complete the construction of the soul S of M. The arguments from this point up to Theorem 8.9 do not depend on any curvature assumptions.

Because all of our arguments use only local (as opposed to total) convexity, we will recall some information from the appendix to Chapter 5. A subset A of M is called strongly convex if for any $q, q' \in A$ there is a unique minimal geodesic segment $\tau_{q,q'}$ from q to q' and $\tau_{q,q'} \subset A$. There exists a positive continuous function $r : M \to [0, \infty)$, the convexity radius, such that any open metric ball $B(p') \subset B_{r(p)}(p)$ is strongly convex. We say that a set $C \subset M$ is *convex* if for any $p \in \operatorname{cl} C$ there is a number $0 < \epsilon(p) < r(p)$ such that $C \cap B_{\epsilon(p)}(p)$ is strongly convex. A totally convex set is of course convex and connected. Notice also that the closure if a convex set is again convex.

Assume C is convex. Let $0 \leq k \leq n$ denote the largest integer such that the collection $\{N_\alpha\}$ of smoothly embedded k-dimensional submanifolds of M which are contained in C is nonempty. Let $N = \bigcup_\alpha N_\alpha$. Let $p \in N$. Then $p \in N_\alpha$ for some α, and since N_α is a smoothly embedded submanifold, there is a neighborhood $U \subset N_\alpha \cap B_{\epsilon(p)/2}(p)$ of p in N_α and a positive $\delta < \frac{1}{2}\epsilon(p)$ such that the exponential map restricted to the set of vectors of length $< \delta$

in the normal bundle of U is a diffeomorphism onto a neighborhood T_δ of p in M. In order to show that N is an embedded submanifold, it will suffice to show that $N \cap T_\delta = U$. In fact, if $q \in (C \cap T_\delta) - U$, and q' is the point of U closest to q, then the minimal geodesic from q to q' is perpendicular to U. Then for a sufficiently small open neighborhood $q' \in U' \subset U$, the minimal geodesic from q to any $q'' \in U'$ intersects U' transversally. It follows that the cone

$$\{\exp(tu) \mid u \in M_q, ||u|| < \epsilon(q), \exp(u) \in U', 0 < t < 1\}$$

is a $(k+1)$-dimensional smooth submanifold of M which is contained in C by convexity. But this contradicts the definition of k. By convexity of C and the existence of T_δ, it is immediate that the submanifold N is totally geodesic. It remains to show that N is connected and $C \subset \overline{N}$. This is an easy consequence of the following lemma.

LEMMA 8.6. *Let C be convex and connected, and let*

$$p \in C \cap \overline{N}, \qquad p' \in B_{\epsilon(p)/4}(p) \cap C, \qquad q \in B_{\epsilon(p)/4}(p) \cap N.$$

Let γ be the normal geodesic in M such that $\gamma|[0,\epsilon]$ is the minimal segment from q to p', where $\epsilon = \rho(q,p')$. Then $\gamma|[0,\epsilon) \subset N$ and hence $p' \in \overline{N}$. If furthermore $p' \notin N$, then $\gamma(s) \notin C$ for all $\epsilon < s < \epsilon + \frac{1}{4}\epsilon(p)$.

PROOF. Let W be sufficiently small $(k-1)$-dimensional hypersurface of $B_{\epsilon(p)/4}(p) \cap N$ through q which is transversal to γ. (See Fig. 8.3.) Let $0 < \hat{\epsilon} < \epsilon + \frac{1}{4}\epsilon(p)$ and $\hat{p} = \gamma(\hat{\epsilon}) \in C$. The cone

$$V = \{\exp(tw) \mid w \in M_{\hat{p}}, ||w|| < \epsilon(p), \exp(w) \in W, 0 < t < 1\}$$

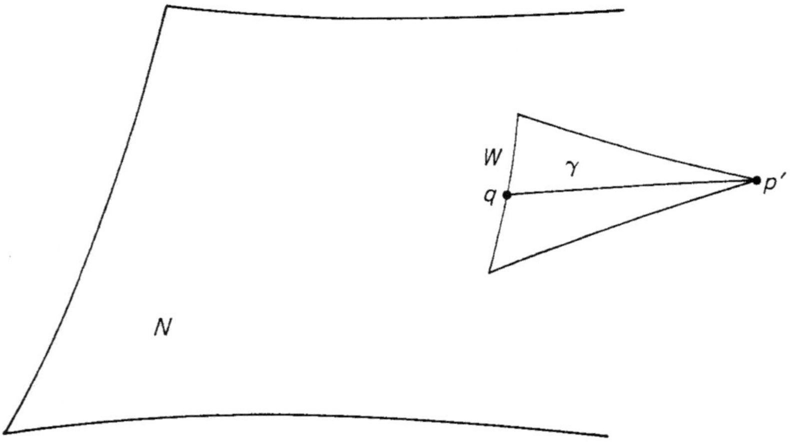

Fig. 8.3.

is a smooth k-dimensional submanifold of M and $V \subset C$, since C is convex. Hence $V \subset N$ by construction of N. In particular, taking $\hat{p} = p'$, we have

$\gamma[0, \epsilon) \subset N$. On the other hand, $\epsilon < \hat{\epsilon} < \frac{1}{4}\epsilon(p)$ would imply $p' \in N$. The lemma follows. $\square$

LEMMA 8.7. *N is connected and $C \subset \overline{N}$.*

PROOF. Let N_0 be a connected component of N. It follows that $C \subset \overline{N}_0$. Otherwise, since C is connected, there exist

$$p \in C \cap \overline{N}_0, \qquad p' \in B_{\epsilon(p)/4}(p) \cap (C - \overline{N}_0), \qquad q \in B_{\epsilon(p)/4}(p) \cap N_0.$$

But $p' \in \overline{N}_0$ by Lemma 8.6. Suppose N_1 is another connected component of N. Since both N_0 and N_1 are dense in C, we must have $N_0 = N_1 = N$. This completes the proof. $\square$

It is now easy to show that a closed onvex set C is an embedded topological submanifold with smooth interior N and possibly nonsmooth boundary (which might be empty). We shall not prove this result because we do not need it. What we shall use instead is the following discussion of tangent cones.

The *tangent cone* C_p at $p \in C$ is defined by the set

$$\{v \mid v \in M_p, \exp_p(tv/||v||) \in N \text{ forsomepositive } t < \epsilon(p)\} \cup \{0\} \subset M_p.$$

Let $\widehat{C}_p$ denote the subspace of M_p generated by C_p. Clearly, if $p \in N$, then $C_p = \widehat{C}_p = N_p$. If, on the other hand, $p \in \partial C = C - N$, then by Lemma 8.7 there exists $q \in B_{\epsilon(p)/4}(p) \cap N$. Letting γ denote the minimal normal geodesic from p to q, by Lemma 8.6 it follows that $\gamma'(0) \in C_p$ but $-\gamma'(0) \notin C_p$. Hence in this case $C_p \neq \widehat{C}_p$. Moreover, if $q' \in B_{\epsilon(p)/4}(p) \cap N$ and if for sufficiently small t, σ_t denotes the minimal normal geodesic from $\gamma(t)$ to q', then

$$\sigma_0'(0) = \lim_{t \to 0} \sigma_t'(0).$$

Since N is totally geodesic, $N_{\gamma(t)}$ is invariant under parallel translation along γ. As usual, letting $P_{[\ ]}$ denote parallel translation along [ ], we have $\widehat{C}_p \subset P_{-\gamma}(N_q)$. However, as in the proof of Lemma 8.6, if we form the geodesic cone at p and a small hypersurface W of N through q and transversal to γ, we see that $C_p - 0$ is *open* in $P_{-\gamma}(N_q)$. Therefore, $\widehat{C}_p = P_{-\gamma}(N_q)$. In particular,

$$\dim C_p = \dim \widehat{C}_p = \dim N.$$

Moreover, if

$$p' \in B_{\epsilon(p)/4}(p) \cap \partial C \text{ and } q \in B_{\epsilon(p)/4}(p) \cap N$$

Lemma 8.7), then for minimal segments $\gamma, \bar{\gamma}$ from p to p' to q, we have

$$\widehat{C}_p = P_{-\gamma \cup \bar{\gamma}}(\widehat{C}_{p'}).$$

It follows that $\widehat{C}_p$ varies continuously with p.

LEMMA 8.8. *Let $C \subset M$ be closed and convex. Suppose that there exist $p \in \partial C, q \in \operatorname{int} C$, and a minimal normal geodesic $\gamma : [0, d] \to C$ from q to p such that $L[\gamma] = \rho(q, \partial C)$. Then $C_p - 0$ is the open half-space*

$$H = \{v \in M_p | \measuredangle(v, -\gamma(d)) < \pi/2\}.$$

PROOF. Take $s < d$ such that $\rho(\gamma(s), p) < \frac{1}{2}\epsilon(p)$. Clearly $\gamma|[s, d]$ realizes the distance from $\gamma(s)$ to ∂C and hence

$$\overline{B_{d-s}(\gamma(s))} \cap \partial C = p.$$

It follows easily that $C_p \supset H$. Conversely, let $q' \in B_{\epsilon(p)}(p) \cap C$ and let $\bar{\gamma}$ denote the minimal geodesic from p to q'. If

$$\measuredangle(\bar{\gamma}'(0), -\gamma'(d)) > \frac{1}{2}\pi,$$

then $-\bar{\gamma}'(0)$ points into $B_{d-s}(\gamma(s))$. Hence, by Lemma 8.6 we would have $p \in \operatorname{int} C$. Since, as we have previously remarked, $C_p - 0$ is open in $\widehat{C}_p$, this suffices to complete the proof.

Any half-space $H \subset \widehat{C}_p$ with the property $C_p \subset \overline{H}$ will be called a supporting half-space for C at $p \in \partial C$. If $p \in \partial C$ and $q \in \operatorname{int} C$ with $\rho(p, q) < \epsilon$ (Lemma 8.7), then the closest point $p' \in \partial C$ to q satisfies $\rho(p, p') < 2\epsilon$. Hence by Lemma 8.8, the set of points having a unique supporting half-space is dense in ∂C. Moreover, if $p \in \partial C$ is arbitrary, and $p_i \to p$ where $p_i \in \partial C$ has supporting half-space H_i, then by local compactness of the unit sphere bundle, we may assume H_i converges to some half-space H. By an argument similar to the discussion above, it follows that H is a supporting half-space for C at p. □

The following theorem is the key to constructing the soul of M. For C a closed and convex set with $\partial C \neq \emptyset$, we set

$$C^a = \{p \in C | \rho(p, \partial C) \geq a\} \text{ and } C^{\max} = \bigcap_{C^a \neq \emptyset} C^a.$$

THEOREM 8.9. *Let M have nonnegative curvature, let C be closed and convex, respectively totally convex, and let $\partial C \neq \emptyset$. Then:*
(1) for any a, C^a is convex, respectively totally convex,
(2) $\dim C^{\max} < \dim C$.

Theorem 8.9 is actually a corollary of the following more general theorem.

THEOREM 8.10. *With the hypothesis of Theorem 8.9, let $\psi : C \to \mathbb{R}$ be defined by $\psi(x) = \rho(x, \partial C)$. Then for any normal geodesic segment γ contained in C, the function $\psi \circ \gamma(t)$ is (weakly) convex, i.e.*

$$\psi \circ \gamma(\alpha t_1 + \beta t_2) \geq \alpha \psi \circ \gamma(t_1) + \beta \psi \circ \gamma(t_2),$$

where $\alpha, \beta \geq 0$, and $\alpha + \beta = 1$. Moreover, suppose

$$\psi \circ \gamma(t) \equiv d$$

is constant on some interval $[a, b]$. *Let* $V(s)$ *denote the parallel field along* $\gamma|[a, b]$ *such that* $V(a) = \dot{\gamma}_a(0)$, *where* γ_a *is any minimal geodesic from* $\gamma(a)$ *to* ∂C. *Then for any* s

$$t \to \exp_{\gamma(s)} tV(s)|[0, d]$$

is a minimal geodesic from $\gamma(s)$ *to* ∂C *and the rectangle*

$$\phi : [a, b] \times [0, d] \to M$$

defined by

$$\phi(s, t) = \exp_{\gamma(s)} tV(s)$$

is flat and totally geodesic.

Theorem 8.10 implies Theorem 8.9 as follows: If $\gamma : [0, d] \to C$ were a normal geodesic segment such that $\gamma(0), \gamma(d) \in C^a$ but $\gamma[0, d] \not\subseteq C^a$, then $\psi \circ \gamma$ would have a strict interior minimum which is impossible for a convex function. This proves (1), and (2) is then trivial.

PROOF OF THEOREM 8.10. For simplicity we will assume that C is totally convex. If C is convex the proof must be slightly modified and we leave this to the reader. The second part of the theorem follows easily from (a) below.

Let $\gamma : [a, b] \to C$ be a normal geodesic. For $a < s_0 < b$ let $\gamma_{s_0} : [0, d] \to C$ be a minimal connection from $\gamma(s_0)$ to ∂C. It will suffice to show that on some interval $(s_0 - \delta, s_0 + \delta)$, $\psi \circ \gamma$ is bounded above by the linear function $d - \cos\alpha(s - s_0)$, where

$$\alpha = \sphericalangle(\gamma'(s_0), \gamma'_{s_0}(0)),$$

for this implies that $\psi \circ \gamma$ is locally the minimum of linear (and therefore convex) functions. The minimum of convex functions is convex. Hence $\psi \circ \gamma$ is locally convex. For the argument, it is only necessary to consider $s > s_0$. Following an idea suggested by the proof of Toponogov's Theorem we distinguish three cases.

(a) $\alpha = \frac{1}{2}\pi$. Let $E(t)$ denote the parallel field along γ_{s_0} generated by $\gamma'(s_0)$. By Rauch II, if follows that there exists $\delta > 0$, such that for $s_0 \leq s \leq s_0 + \delta$, the curve

$$\gamma_s(t) = \exp_{\gamma_{s_0}}(s - s')E(t)$$

has length $\leq d = r - \cos\frac{1}{2}\pi$. If equality holds for some s', then it must also hold for all smaller s'', as follows easily from the proof of Rauch II. The curves

$$s \to \exp_{\gamma_{s(t)}} sE(t)$$

must be geodesics, since they are minimal and from the proof of Rauch II the totally geodesic rectangle which they therefore determine must be flat. The function

$$\gamma_s(t) : [s_0, s'] \times [0, d] \to C$$

defines a flat totally geodesic rectangle. Clearly $\psi \circ \gamma \equiv d$ unless for some fixed s the curve $\gamma_s(t) \subset \operatorname{int} C$ for $0 \le t \le d$. On the other hand, by Lemma 8.8, this is impossible for $t = d$.

(b) $\alpha > \frac{1}{2}\pi$. In this case let $E(0)$ be the convex combination of $\gamma'_{s_0}(0)$ and $\gamma'(s_0)$ which is perpendicular to $\gamma'_{s_0}(0)$. Constructing a family $\gamma_s(t)$ as above, it follows as in (a) that for small δ and $s_0 \le s \le s_0 + \delta$, that

$$\rho(\exp_{\gamma(s_0)} \cos\left(\alpha - \frac{1}{2}\pi\right)(s - s_0)E(0), \partial C) \le d.$$

But by Rauch I,

$$\begin{aligned}\rho(\gamma(s), \exp_{\gamma(s_0)} \cos\left(\alpha - \frac{1}{2}\pi\right)(s - s_0)E(0)) &\le (s - s_0)\sin\left(\alpha - \frac{1}{2}\pi\right)\\ &= -(s - s_0)\cos\alpha.\end{aligned}$$

So by the triangle inequality,

$$\rho(\gamma(s), \partial C) \le d - (s - s_0)\cos\alpha.$$

(c) $\alpha < \frac{1}{2}\pi$. Let a_s be the minimal segment from γ_{s_0} to $\gamma(s)$, where $a_s(0) = \gamma_{s_0}(t_s)$. Then $a'_s(0) \perp \gamma'(t_s)$. As above,

$$\rho(\gamma(s), \partial C) \le d - t_s.$$

By Rauch I and the law of cosines,

$$(\rho(\gamma_{s_0}(t_s), \gamma(s)))^2 \le t_s^2 + (s - s_0)^2 - 2t_s s\cos\alpha.$$

Since the angle at $\gamma_{s_0}(t_s)$ is a right angle, we also have

$$(s - s_0)^2 \le (\rho(\gamma_{s_0}(t_s), \gamma(s)))^2 + t_s^2.$$

Combining these two inequalities yields

$$2t_s(s - s_0)\cos\alpha \le 2t_s^2$$

or

$$(s - s_0)\cos\alpha \le t_s.$$

Thus again

$$\rho(\gamma(s), \partial C) \le d - \cos\alpha(s - s_0).$$

□

Now choose a compact t.c.s C as in Proposition 8.5 such that $\partial C \ne \emptyset$. By iterating Theorem 8.9, we may construct a flag of compact t.c.s., where each set consists of all points at maximum distance from the boundary of the preceding set and is of lower dimension. Thus we have by induction:

THEOREM 8.11. *M contains a compact totally geodesic submanifold S without boundary which is totally convex, $0 \le \dim S < \dim M$. In particular, S has nonnegative curvature.*

A manifold S constructed as above will be called a *soul* of M. The construction of S depended on the choice of some arbitrary point $p \in M$. A different choice of p will sometimes lead to a different soul. S may sometimes consist of a single point.

In a sense Theorem 8.11 is a rigid version of the following result, which was proved first (see Gromoll and Meyer [1969]), and without resort to Toponogov's Theorem. Still earlier results for the two-dimensional case are due to Cohn-Vossen [1935], [1936].

COROLLARY 8.12. *If $K_M > 0$, then a soul of M is a point.*

PROOF. This is clear from Theorem 8.10. □

It will be convenient to change our notation slightly by reindexing the expanding and contracting families of t.c.s. we have constructed. Let C_0 be the t.c.s. of Theorem 8.9. In case $\dim C_0 = \dim M$ let us now agree to denote $C_0^{\max} = C_0^{a_0}$ by C_0. Then what was previously denoted by C^t becomes C_{t+a_0} and C_0^a becomes C_{a_0-a}. Conclusions (1),(2) of Theorem 8.9 hold for the family as reindexed. In addition, with our new notation

$$\dim C_0 < \dim M, \qquad \dim C_t = \dim M, \qquad t > 0.$$

Moreover, we have the flag of t.c.s.

$$C_0 = C(1) \supset \cdots \supset C(k) = S,$$

where $C(i+1) = C(i)^{\max}$.

EXAMPLE 8.13 (continued). The (unique) soul of the paraboloid is the vertex.

EXAMPLE 8.14 (continued). Any circle going around the cylinder perpendicular to the axis is a soul.

EXAMPLE 8.15. The tangent bundle $T(S^2)$ of the 2-sphere admits a complete metric of nonnegative curvature.

This example is especially interesting because the metric is not locally a product even though the soul (the zero section) is not a point. The metric may be described as follows: Take the $T(S^2)$ with its standard connection coming from the metric of constant curvature on S^2. At each point v in a fibre F there is a horizontal complement H to the fibre determined by the connection. Let H be perpendicular to F and define the metric on H by making the projection $\pi : T(S^2) \to S^2$ an isometry (with the standard metric on S^2). Instead of the usual flat metric $(\mathrm{d}r^2 + r^2\mathrm{d}\theta^2)$ on the fibres, use the metric,

$$\mathrm{d}r^2 + \frac{r^2}{1+r^2}\mathrm{d}\theta^2.$$

A more direct way of viewing the metric on $T(S^2)$ arises from Example 3.43. Take the Riemannian product $S^3 \times \mathbb{R}^2$, where S^3 has its standard metric of constant curvature which is bi-invariant with respect to its Lie-group structure. Let

$$\phi \to e^{\phi x} = e^{(2\pi+\phi)x}$$

denote a (closed) 1-parameter subgroup of S^3 and let S^1 act by isometries on $S^3 \times \mathbb{R}^2$ by

$$\phi(g, r, \theta) = (ge^{\phi x}, r, \theta + \phi).$$

This gives rise to a fibration

$$\begin{array}{ccc} S^1 & \longrightarrow & S^3 \times \mathbb{R}^2 \\ & & \downarrow \pi \\ & & T(S^2) \end{array}$$

Since S^1 acts by isometries, we get an induced metric on the quotient space $T(S^2)$ by setting $||x|| = ||\pi^{-1}(x)||$, where $\pi^{-1}(x)$ is any inverse image of x which is perpendicular to the fibre S^1. π becomes a Riemannian submersion and the formula of O'Neill implies that the curvature of $T(S^2)$ is nonnegative. One can check that the metric on $T(S^2)$ is the one we have described above.

We will now give an indication of the proof that the inclusion $i : S \to M$ is a homotopy equivalence.

THEOREM 8.16. *Let C be a compact totally convex set in M, $\partial C = \emptyset$. Then the inclusion $C \subset M$ is a homotopy equivalence.*

PROOF. We consider the space Ω_C of piecewise smooth curves $c : [0, 1] \to M$ such that $c(0), c(1) \in C$. As in Chapter 4, one has the space Ω_C^a of curves of energy $\leq a$ which may be replaced by a finite-dimensional manifold of broken geodesics. Since C is totally convex, the constant geodesics

$$\gamma : [0, 1] \to p \in C$$

are the only critical points of the energy function on Ω_C^a and these form a submanifold diffeomorphic to C. Hence using the trajectories of the gradient of E there is a strong deformation retract of Ω_C onto C. Let D^k denote the k-disc. Now a map

$$(D^k, \partial D^k) \to (M, C)$$

induces in a standard fashion a map

$$(D^{k-1}, \partial D^{k-1}) \to (\Omega_C, C)$$

and vice versa, for $k \geq 1$. Therefore $\pi_k(M, C)$ is trivial for all $k \geq 1$. The theorem now follows from the homotopy sequence of a pair and the Whitehead Theorem (Spanier [1966]). □

Theorems 8.11 and 8.16 imply that if $K_M > 0$, then M is contractible.

We are now going to discuss the special cases where codim $< S = 1$ or dim $S = 1$. We begin with a slight generalization of Lemma 6.2 which is proved in just the same way. The proof will be omitted.

LEMMA 8.17. *Let D be a compact subset of a closed t.c.s. C in an arbitrary complete Riemannian manifold M. If the function $q \to \rho(q, D)$ assumes a (relative) maximum at $p \in \operatorname{int} C$, then for any $v \in C_p$ there is a minimal geodesic $\sigma : [0, d] \to M$ from p to D such that $\langle v, \dot\sigma(0)\rangle \geq 0$.*

THEOREM 8.18. *Let $S \subset M$ be a soul and suppose* $\dim S = \dim M - 1$. *Let $\nu(S)$ be the normal bundle of S. Then* $\exp_S|\nu(S)$ *is an isometry between $\nu(S)$ with its standard (flat) bundle metric and M.*

PROOF. By Proposition 8.5, $M = \bigcup_t C_t$, where C_t is totally convex and $S = C_1^{\max}$. Now for any $p \in S$ and $t > 0$ there is a minimal connection σ to ∂C_t. However, since all points of S are equidistant from ∂C_t we have $\dot\sigma(0) \in \nu(S)$. But by Lemma 8.17 it then follows that $-\sigma$ is also a minimal connection to ∂C_t for any t. By the argument of Theorem 8.10 it follows that $\exp_S|\nu(S)$ is a local isometry (where $\nu(S)$ has its induced metric) and hence a covering. Since S is totally convex it follows that each $p \in S$ has just one inverse image and hence that $\exp_S|\nu(S)$ is a global isometry. □

To handle the case $\dim S = 1$ we use the following Splitting Theorem which is actually valid under the weaker hypothesis $\mathrm{Ric}_M \geq 0$. The proof for that case is more difficult and relies on results from analysis outside the scope of this book.

A *line* is a geodesic $\gamma : (-\infty, \infty) \to M$, each finite segment of which realizes the distance between its endpoints. If M has positive curvature, an argument very much like Theorem 1.31 shows that M contains no lines. Thus the result which follows is a rigidity theorem.

THEOREM 8.19 (Cheeger and Gromoll [1972], Toponogov [1964]). *M may be written uniquely as the isometric product $\overline{M} \times \mathbb{R}^k$, where $\overline{M}$ contains no lines and $\mathbb{R}^k$ has its standard flat metric.*

PROOF. By induction, it suffices to show that if M contains a line σ, then M is isometric to $\overline{M} \times \mathbb{R}$. Now it follows from the fact that σ is a line that B_{σ_a} and $B_{-\sigma_a}$ are at distance precisely $2a$. By Proposition 8.5 and Lemma 8.8, any geodesic $\tau : (-\infty, \infty) \to M$ such that

$$\tau(0) = \sigma(0), \text{ and } \tau'(0) \perp \sigma'(0)$$

does not meet $\operatorname{int}((B_{\sigma_0})')$. The prime $'$ denotes set-theoretic complement. Hence by Theorem 8.10, $\rho(\tau(t), \partial B_{-\sigma_a}) = a$ for all t; similarly $\rho(\tau(t), \partial B_{\sigma_a}) = a$. Suppose

$$\rho(q, \partial((B_{-\sigma_a})' \cap (B_{\sigma_a})')) \geq a$$

for some $q \in (B_{-\sigma_a})' \cap (B_{\sigma_a})'$ and let γ be a geodesic from $\sigma(0)$ to q. Then by Proposition 8.5,

$$\gamma \subset (B_{-\sigma_a})' \cap (B_{\sigma_a})',$$

which implies by Lemma 8.8 that $\gamma'(0) \perp \sigma'(0)$. Then by the above

$$\rho(q, \partial((B_{-\sigma_a})' \cap (B_{\sigma_0})')) = a,$$

and therefore

$$((B_{-\sigma_a})' \cap (B_{\sigma_a})')^{\max} = \exp_{\sigma(0)}\nu(\sigma).$$

Furthermore, $\exp_{\sigma(0)} \nu(\sigma)$ has no boundary. If q were a boundary point and $q = \exp_{\sigma(0)} t_0 v$, then by Lemma 8.6 we would have for $t > t_0$ that

$$\exp_{\sigma(0)} tv \notin \exp_{\sigma(0)}\nu(\sigma),$$

which is ridiculous. The proof may now be completed as in Theorem 8.18 □

THEOREM 8.20. *If* $\dim S = 1$ *(in which case* S *is a circle), then* M *is a locally isometrically trivial bundle over* S.

PROOF. Let $\rho : \widetilde{M} \to M$ denote the natural projection of the universal covering space $\widetilde{M}$ on M. Since ρ is a local isometry it follows easily that $\rho^{-1}(S)$ is totally convex. However, since M has the homotopy type of S^1, we have $\pi(M) \cong \mathbb{Z}$ and hence ρ is an infinite covering. Thus $\rho^{-1}(S)$ is a noncompact 1-dimensional totally convex submanifold of $\widetilde{M}$. In other words, $\rho^{-1}(S)$ is a line. Let $\widehat{M} \times \mathbb{R}$ be the corresponding splitting of $\widetilde{M}$ guaranteed by Theorem 8.19. Let $\pi_2 : \widehat{M} \times \mathbb{R} \to \mathbb{R}$ be the projection. It is easy to see that π_2 commutes with covering transformations and hence induces $\pi : M \to S$. Further, it follows easily that if I_1, I_2 form a covering of $S^1 = S$ by open intervals, then $\pi^{-1}(I_i)$ is isometric to $I_i \times \widehat{M}$ in such a way that on $I_1 \cap I_2$ the isometric product structure is preserved. □

We now apply Theorem 8.19 to study the fundamental group of M. As we know, if $K_M > 0$, then $\pi_1(M)$ is finite. We should then expect that if $K_M \geq 0$, $\pi_1(M)$ can be infinite only under very special circumstances.

DEFINITION 8.21. Suppose that M is a Hausdorff space and G is a group which acts on M. We say that G acts *properly discontinuously* if for each $p \in M$ there is a neighborhood $U \ni p$ such that $U \cap g(U) \neq \emptyset$ for at most finitely many $g \in G$. G is said to act uniformly if the orbit space M/G is compact in the identification topology.

If G acts properly discontinuously, then M/G is Hausdorff. A group G of isometries of $\mathbb{R}^k$ which acts properly discontinuously and uniformly is called crystallographic group. The structure of crystallographic groups is given in the following theorem. The reader is referred to Wolf [1966] for a proof. If each element is fixed point free, G is said to be a Bieberbach group.

THEOREM 8.22 (Bieberbach). *Let* G *be a crystallographic group acting on* $\mathbb{R}^k$. *Then* G *contains a normal free abelian subgroup of rank* k*, the group of translations, which is of finite index.* k *is called the rank of* G.

Let G be a properly discontinuous group of isometries of $\mathbb{R}^n$ such that each element is fixed point free. Then $\mathbb{R}^n/G$ is in a natural way a flat Riemannian manifold which might not be compact. However, our main result shows the existence of a flat compact soul S of the homotopy type of M. It follows that G is a Bieberbach group on $\mathbb{R}^k$ with $k = \dim S \leq n$.

In view of Theorem 8.22, the following result provides a good description of the fundamental group in case $K_M \geq 0$.

THEOREM 8.23. *Let M be a complete manifold of nonnegative curvature. Then there exists a finite normal subgroup $\Phi \subset \pi_1(M)$ such that $\pi_1(M)/\Phi$ is a Bieberbach group.*

PROOF. By Theorem 8.11 and Theorem 8.16 we may assume M to be compact. Let $\widetilde{M} \cong \overline{M} \times \mathbb{R}^k$ be the splitting of the universal covering which is guaranteed by Theorem 8.16. The main point is to show that $\overline{M}$ is compact. We identify $\pi = \pi_1(M)$ with the group of isometric covering transformations of $\widetilde{M}$.

(1) π preserves the splitting $\overline{M} \times \mathbb{R}^k$. Hence any element of π is of the form $(x, y) \to (f(x), g(y))$.

(2) There exists d such that for any $p, q \in \widetilde{M}$, we have $\rho(\pi(p), q) < d$. In fact let d be the diameter of M. Let

$$\gamma : [0, l] \to M, \qquad l < d,$$

be a minimal geodesic from $\sigma(q)$ to $\sigma(p)$, where σ denotes the covering projection. Let $\tilde{\gamma} : [0, l] \to \widetilde{M}$ be such that $\tilde{\gamma}(0) = q$ and $d\sigma(\tilde{\gamma}'(0)) = \gamma'(0)$. Since σ is a local isometry, it commutes with exponential maps. Hence $\sigma(\tilde{\gamma}(l)) = \sigma(p)$, which means $\tilde{\gamma}(l) \in \pi(p)$.

(3) Let π_1, π_2 denote the projections of $\overline{M} \times \mathbb{R}^k$ on $\overline{M}$ and $\mathbb{R}^k$ respectively. By (1), we have well-defined groups of isometries $\pi_1 \circ \pi$ and $\pi_2 \circ \pi$. Now since π_1 and π_2 are distance decreasing, it follows from (2) that $\pi_1 \circ \pi$ and $\pi_2 \circ \pi$ act uniformly.

(4) If $\overline{M}$ is not compact there exists a sequence of points p_i such that $\rho(p_0, p_i) \to \infty$. Let $\gamma_i : [0, d_i] \to \overline{M}$ be a sequence of minimal normal geodesics from p_0 to p_i. Then by (3) there exist $g_i \in \pi_1 \circ \pi$ such that

$$\rho(g_i^{-1}(\gamma_i(d_i/2)), p_0) < d.$$

Hence we may choose a subsequence such that

$$g_i^{-1}(\gamma_i(d_i.2)) \to p, \qquad dg_i^{-1}(\gamma_i'(d_i/2)) \to v \in \overline{M}_p.$$

Then if $\gamma : (-\infty, \infty) \to \overline{M}$ such that $\gamma'(0) = v$, then γ is a line. This is a contradiction.

(5) The kernel Φ of π_2 consists of elements of the form $(f, 1)$. Since π is a group of covering transformations there exists $\epsilon > 0$ such that for all $p \in M$ and $g \in \pi$, we have $\rho(g(p), p) > \epsilon$. Since $\overline{M}$ is compact, it follows that Φ is finite.

(6) Similarly it follows that $\pi_2 \circ \pi \cong \pi/\Phi$ is discrete. This suffices to complete the proof. □

We will return briefly to the study of the global behavior of geodesics which was initiated in Chapter 2. The discussion will depend essentially on Proposition 8.5, which was based on Toponogov's Theorem. C_t will denote a filtration of M by t.c.s. as reindexed after Corollary 8.12.

THEOREM 8.24. *Let M be complete noncompact and have nonnegative curvature.*

(1) If $\gamma : (-\infty, \infty) \to M$ is a geodesic and $\gamma \subset C$ for some compact set C, then $\gamma \subset \partial C^a$ for some $a > 0$ or $\gamma \subset C_0$.

(2) Any geodesic intersecting a soul S transversally goes to infinity in both directions.

(3) If $\sigma, \tau : [0, \infty) \to M$ are geodesics such that

$$\sigma(0) = \tau(0), \qquad \langle \sigma'(0), \tau'(0) \rangle > 0$$

and σ is a ray, then τ goes to infinity.

(4) If σ, τ are as in (3) except $\langle \sigma'(0), \tau'(0) \rangle = 0$ and τ does not go to infinity, then $\tau \subset \partial H_\sigma$. If V denotes the parallel field along τ generated by σ', then

$$[0, \infty) \times [0, \infty) \to \exp_{\tau(t)} sV(t)$$

defines a flat (immersed) totally geodesic rectangle.

(5) If M is not contractible and C is a compact t.c.s., then through each point $p \in C$ there is at least one geodesic $\gamma : [0, \infty) \to M$ which stays in C.

(6) If $K_M > 0$, then any geodesic $\gamma : [0, \infty) \to M$ goes to infinity.

Item (4) describes the rigidity phenomenon mentioned in Chapter 2; (2), (3) and half of (1) are weakened generalizations of (6). The other half of (1) and (4) are rigid versions of (6). (5) is in the nature of a converse to (6).

PROOF. (1) If $\gamma : [0, \infty) \to M$ does not go to infinity, let

$$a = \inf_{\gamma \subset C_t} t.$$

We may suppose $a > 0$. By Theorem 8.10,

$$\rho(\gamma(t), \partial C^a) : [0, \infty) \to \mathbb{R}^+$$

is a convex function of t which is bounded below with greatest lower bound 0. Hence $\rho(\gamma(t), \partial C^a) \equiv 0$. (In case $a = 0$, a further refinement based on Theorem 8.10 is obviously possible.)

(2) Let

$$C_0 = C(1) \supset \cdots \supset C(k) = S$$

be the flag of t.c.s. described after Corollary 8.12. Since γ meets S transversally (at say $\gamma(0)$), we have

$$\gamma|_{[0, \pm\infty)} \not\subseteq S = C(k-1)^{\max}.$$

Hence as in (1), $\gamma \not\subseteq C(k-1)$ and (2) follows by induction.

(3) Let C_t denote the filtration of M by t.c.s. as above obtained by doing the basic construction at $p = \sigma(0)$. Suppose $\sigma(0) \in C_a$. By Proposition 8.5, $\sigma|[0, \epsilon]$ is a minimal connection from $\sigma(0)$ to $\partial C_{a+\epsilon}$. Hence $N = \sigma'(\epsilon)^\perp$ is the supporting hyperplane for $C_{a+\epsilon}$ at $\sigma(\epsilon)$. Then $\exp_{\sigma(\epsilon)} N$ lies outside $C_{a+\epsilon}$. It follows from the first variation formula that $\rho(\tau(t), \partial C_{a+\epsilon})$ is not constant. Hence by (1), τ goes to infinity.

(4) This follows by combining the argument of (3) above with the second part of the proof of Theorem 8.10.

(5) If M is not contractible, then by Milnor [1963] $\Omega(M)$ has nonzero homology in infinitely many dimensions. Then if $q \in C$ is not conjugate to p along any geodesic, it follows from Theorem 4.12 that there is a sequence $\{\gamma_i\}$ of geodesics from p to q whose lengths go to infinity. Let v be an accumulation point for $\{\gamma_i'(0)\}$ and take γ such that $\gamma'(0) = v$.

(6) Suppose $\gamma \subset C$. Then since $\rho(\gamma(t), \partial C_a)$ is a convex function which is bounded below, it must be an increasing function which is asymptotic to some constant k. By compactness we can choose a subsequence of $\{\gamma'(i)\}$ converging to some tangent vector v. Then if σ is the geodesic such that $\sigma'(0) = v$, it follows easily that $\rho(\gamma(t), \partial C_a) \equiv k$. By Theorem 8.10, C contains a flat totally geodesic rectangle which contradicts $K_M > 0$. □

CHAPTER 9

Compact Manifolds of Nonpositive Curvature

Manifolds of nonpositive curvature arise naturally in many contexts. For example, it is a well-known fact that a compact orientable 2-dimensional manifold of genus $g > 1$ admits a $(6g - 6)$ dimensional family of metrics of constant curvature -1. For higher-dimensional examples we refer to Borel [1963].

As we have already seen in Corollary 1.40, the universal covering space $\widetilde{M}$ of a complete manifold of nonpositive curvature is diffeomorphic to Euclidean space. Hence $\pi_i(M) = 0$ for $i > 1$, and M is a so called $K(\pi, 1)$ space. In particular, the homotopy type of M is determined by $\pi = \pi_1(M)$. We are going to give a number of results on the relation between the geometric structure of M and the algebraic structure of its fundamental group in case M is compact. The guiding principle is that $\pi_1(M)$ is big if the curvature of M is strictly negative. For example, a result of Milnor [1968] says that $\pi_1(M)$ has "exponential growth" in this case. Moreover, according to the theorem of Preismann (Corollary 9.9) abelian subgroups of $\pi_1(M)$ must be infinite cyclic.

The results of the present chapter are basically rigidity theorems which measure the extent to which the Preismann Theorem fails in the case that K_M is non-positive. Our two main results are due to Gromoll and Wolf [1971] and in a very similar form to Lawson and Yau [1972].

We begin by recalling that a group Σ is said to be *solvable* if there exists a finite sequence of subgroups

$$\Sigma = \Sigma^0 \supset \Sigma^1 \supset \cdots \supset \Sigma^n = e$$

such that Σ^{i+1} is normal in Σ^i and Σ^i/Σ^{i+1} is abelian. Also we define An isometry $g : M \to M$ to be *semi-simple* if the *displacement function* $\delta_g(p) = \rho(g(p), p)$ assumes a minimum $m_g \geq 0$.

THEOREM 9.1. *Let M be closed totally convex subset of a simply connected manifold of nonpositive curvature and let Γ be a properly discontinuous group of semi-simple isometries of M. Let Σ be a solvable subgroup of Γ. Then M contains a flat totally geodesic submanifold E, which is isometric to $\mathbb{R}^k$ and invariant under Σ, such that:*

(1) Σ acts with finite kernel Φ on E;

(2) E/Σ is compact. In particular, Σ is finitely generated and Σ/Φ is a crystallographic group of rank k.

In particular, (1) and (2) hold if M is the universal covering space of a compact manifold and Γ the group of covering transformations.

Theorem 9.1 should be compared to Theorem 8.23, which gives a somewhat analogous result in the case $K_M \geq 0$. We have stated Theorem 9.1 with the hypothesis that M is a totally convex set (rather than simply a complete manifold) because it is necessary to work in this degree of generality to carry out various induction steps in the proof. Quite naturally, we will need the theory of convex sets as developed in Chapter 8. In particular, implicitly we are always assuming that a t.c.s. is a submanifold with totally geodesic interior.

We will also use without proof, the de Rham Decomposition Theorem. This theorem states that if the tangent bundle $T(M)$ of a complete simply connected Riemannian manifold M splits as a direct sum $T(M) = \xi \oplus \eta$ in a manner invariant under parallel translation, then M splits isometrically as a product $M = M_1 \times M_2$, where the tangent spaces to the factors give the splitting $\xi \oplus \eta$. For a proof see Kobayashi and Nomizu [1963, 1969].

Here is an outline of the proof of Theorem 9.1, the details of which will require five lemmas. The basic geometric ingredient is the fact that if $K_M \leq 0$, the displacement function δ_g of an isometry is concave. Thus the set

$$C_g^a = \{p \in M | \delta_g(p) \leq a\}$$

is convex. Also the minimum set of δ_g, which we shall call C_g, is convex. If g has infinite order, then C_g will be shown to split isometrically as $D \times \mathbb{R}$, where each $d \times \mathbb{R}$ is a line invariant under g. Now if Σ is abelian, by induction $\bigcap_{s \in \Sigma} C_s$ is nonvoid and splits as $D' \times \mathbb{R}^k$. We may then take $E = d \times \mathbb{R}^k$. The extension from abelian to solvable proceeds by another induction on the length of the series, $\Sigma^0 \supset \cdots \supset \Sigma^n$.

Our first lemma collects some results about isometric group actions and does not involve any assumptions about sectional curvature. (5) was already used in Theorem 8.23.

LEMMA 9.2. *Let M be a totally convex subset of a complete Riemannian manifold and let Γ, Σ be the groups of isometries of M.*

(1) If Γ is properly discontinuous and uniform, then for each $g \in \Gamma$, δ_g assumes its minimum, i.e. Γ is semi-simple.

(2) If Σ is a normal subgroup of Γ, then $C_\Sigma = \bigcap_{s \in \Sigma} C_s$ is invariant under Γ.

(3) If Γ acts uniformly on M, then it acts uniformly on any Γ invariant subset of M.

(4) Suppose M splits isometrically as $D \times \mathbb{R}$. Let g be an isometry which preserves the splitting, and acts by translation on $\mathbb{R}$. If g is semi-simple, then the projected action of g D is also semi-simple.

(5) Let M split isometrically as $D_1 \times D_2$, and let Σ be a properly discontinuous group of isometries which preserves the splitting. Suppose A,

the kernel of the projection on D_1*, acts uniformly on* D_2*. Then* Σ/A *acts properly discontinuously on* D_1*.*

(6) If Γ *acts uniformly and* g *commutes with every element of* Γ*, then* δ_g *is bounded.*

(7) Let Σ *act properly discontinuously on* M*. Let* g *be an isometry of* M*. Then* g *induces an isometry of* M/Σ *if and only if* $g\Sigma g^{-1} = \Sigma$*,* g *induces the identity on* M/Σ *if and only if* $g \in \Sigma$*.*

PROOF. (1) Fix $p \in M$, and let $\{p_i\}$ be such that

$$\delta_g(p_i) \to l = \inf \delta_g.$$

Since Γ is uniform, there exist $d > 0$ and $\{g_i\}, \{q_i\}$ such that

$$p_i = g_i(q_i)$$

with $q_i \in B_d(p)$. Now for any m,

$$\begin{aligned}\delta_g(m) = \rho(g(m), m) &= \rho(g_i g_i^{-1} g g_i g_i^{-1}(m), g_i g_i^{-1}(m)) \\ &= \rho(g_i^{-1} g g_i g_i^{-1}(m), g_i^{-1}(m)) = \delta_{g_i^{-1} g g_i}(g_i^{-1}(m)).\end{aligned}$$

Hence

$$\inf \delta_g = \inf \delta_{g_i^{-1} g g_i}$$

and in particular

$$\delta_g(p_i) = \delta_{g_i^{-1} g g_i}(q_i).$$

Thus

$$\delta_{g_i^{-1} g g_i}(p) \leq l + 2d$$

for i sufficiently big, and the sequence $\{g_i^{-1} g g_i(p)\}$ has an accumulation point. But this contradicts the proper discontinuity of Γ unless the set $\{g_i^{-1} g g_i\}$ is finite. Hence we can find g_0 and $\bar{g}$ defined to be $g_0^{-1} g g_0$ such that $g_i^{-1} g g_i = \bar{g}$ for infinitely many i. Let us call the set of such i $\mathcal{I}$ Then for $i \in \mathcal{I}$ we have $\delta_{\bar{g}}(q_i) \to l$. Let q be an accumulation point of $\{q_i | i \in \mathcal{I}\}$. Then $q \in C_{\bar{g}} \neq \emptyset$ and $g_0(q) \in C_g$.

(2) As above,

$$\delta_s(gq) = \delta_{g^{-1} s g}(q).$$

Then

$$g(C_\Sigma) = g(\bigcap_{s \in \Sigma} C_s) = \bigcap_{s \in \Sigma} g(C_s) = \bigcap_{s \in \Sigma} C_{g^{-1} s g} = \bigcap_{s \in \Sigma} C_s = C_\Sigma.$$

(3) Let C be a Γ-invariant subset, and K a compact subset of M such that $\Gamma(K) = M$. Then $C \cap K$ is a relatively compact subset of C and

$$C = C \cap M = C \cap \Gamma(K) = \Gamma(C) \cap \Gamma(K) = (C \cap K).$$

(4) Let g act by

$$(p, q) \to (f(p), t + q).$$

Then

$$\delta_g(p, q) = (\rho(p, f(p))^2 + t^2)^{1/2}.$$

Clearly if δ_g achieves its minimum at (p_0, q_0), then $\rho(p, f(p))$ achieves its minimum at $p = p_0$.

(5) Let a typical element s of Σ act by

$$(p, q) \to (f(p), g(q)).$$

If Σ/A does not act properly discontinuously, we can find a sequence $s_i = (f_i, g_i)$ and a point p_0 such that $f_i(p_0)$ has a limit point. Now for any q_0 there exists a compact set $K \subset D_1$ and elements $A \ni \bar{s}_i = (1, h_i)$ such that $h_i^{-1} g_i(q_0) \in K$. Then $(\bar{s}_i)^{-1} s_i$ is a sequence of distinct elements such that $(\bar{s}_i)^{-1} s_i(p_0, q_0)$ has a limit point. In this case Σ could not act properly discontinuously.

(6) Let K be a compact set such that $\Gamma(K) = M$. Suppose p_i is a sequence of points such that $\delta_g(p_i) \to \sup \delta_g$. Then there exist a sequence $s_i \in \Gamma$ such that $s_i^{-1}(p_i) \in K$. But

$$\delta_g(p_i) = \delta_{s_i g s_i^{-1}}(s_i^{-1}(p)) = \delta_g(s_i^{-1}(p)).$$

Thus

$$\sup_M \delta_g = \sup_K \delta_g < \infty.$$

(7) g induces an isometry of M/Σ if and only if it maps orbits onto orbits. In other words, for all p,

$$g s_1(p) = s_2 g(p).$$

Since orbits are discrete, it follows by continuity that for fixed s_1, the associated s_2 is independent of p. Hence $g s_1 = s_2 g$ which is equivalent to $g s_1 g^{-1} = s_2$. g induces the identity if and only if for all p,

$$g s_1(p) = s_2(p).$$

Then

$$g s_1 = s_2, \qquad g = s_2 s_1^{-1}.$$

□

Our next two lemmas, while still of a rather general nature, bring in the non-positive curvature of M. Throughout the next four lemmas, M is assumed to be a totally convex subset of a complete simply connected manifold of non-positive curvature.

LEMMA 9.3. *Let g be an isometry of M. Then $\delta_g|\gamma$ is concave, for any geodesic γ.*

PROOF. Let γ be parameterized by arclength, and $\sigma_s(t) : [0, l] \to M$ the unique minimal normal geodesic (non-positive curvature) from $\gamma(s)$ to $g(\gamma(s))$. Then $g(\gamma(s))$ is also a normal geodesic. If $\gamma(s) \neq g(\gamma(s))$, we use the second variation formula to calculate

$$L''[\sigma_s(t)] = \frac{d^2}{ds^2} \delta_g|\gamma(s).$$

Let V denote the variation vector field and let $T = \sigma_s'$. Then

$$\begin{aligned} L''[\sigma_s(t)] &= \langle \nabla_V V, T \rangle \Big|_0^l + \int_0^l \langle \nabla_T V, \nabla_T V \rangle - \langle R(V,T)T, V \rangle - \langle \nabla_T V, T \rangle^2 \\ &= \int_0^l -K(V,T)||V \wedge T||^2 + ||\nabla_T V \wedge T||^2 \\ &\geq 0, \end{aligned}$$

since

$$||V \wedge T||^2 = \langle V, V \rangle \langle T, T \rangle - \langle V, T \rangle^2 \geq 0.$$

Since the second derivative is nonnegative, $\delta_g|\gamma(t)$ is concave. If $\gamma(s) = g(\gamma(s))$, then at its single point of nondifferentiability, δ_g has a minimum. Therefore δ_g is concave in this case as well. □

We say that normal geodesic segments $(\gamma_0, \gamma_1, \gamma_2, \gamma_3)$ determine a *geodesic quadrilateral* if

$$\gamma_i(l_i) = \gamma_{i+1}(0) \bmod 4$$

We let $\alpha_i = \cos^{-1}\langle -\gamma_{i-1}'(l_{i-1}), \gamma_i'(0) \rangle$ where $\cos^{-1}$ has range $[0, \pi]$.

LEMMA 9.4. *Let $Q = (\gamma_0, \gamma_1, \gamma_2, \gamma_3)$ determine a geodesic quadrilateral in M, and assume $\sum \alpha_i \geq 2\pi$. Then for any vertex $\gamma_i(0), \exp_{\gamma_i(0)}^{-1}(\gamma_{i+1})$ and $\exp_{\gamma_i(0)}^{-1}(\gamma_{i+2})$ are contained in the 2-dimensional subspace $S \subset M_{\gamma_i(0)}$ spanned by $\gamma_i'(0)$ and $-\gamma_{i-1}'(l_{i-1})$. $\exp_{\gamma_i(0)}^{-1}(Q)$ bounds a region R in S and $\exp_{\gamma_i(0)}|R$ is an isometry. Thus Q bounds a flat totally geodesic submanifold, and $\sum \alpha_i = 2\pi$.*

PROOF. Let $\sigma : [0, d] \to M$ be the minimal normal geodesic from $\gamma_1(l_1)$ to $\gamma_0(0)$. Let

$$\begin{aligned} &\sphericalangle(\gamma_0'(0), -\sigma'(d)) = \beta, \quad \sphericalangle(-\gamma_3'(l_3), -\sigma'(d)) = \hat{\beta}, \\ &\sphericalangle(-\gamma_1'(l_3), \sigma'(0)) = \delta, \quad \sphericalangle(\gamma_2'(0), \sigma'(0)) = \hat{\delta}. \end{aligned}$$

Then $\beta + \hat{\beta} \geq \alpha_0$ and $\delta + \hat{\delta} \geq \alpha_2$. Let $(\bar{\sigma}, \bar{\gamma}_0, \beta)$ determine a hinge in $\mathbb{R}^n$ with

$$L[\sigma] = L[\bar{\sigma}], \qquad L[\gamma_0] = L[\bar{\gamma}_0]$$

and let $I : M_{\gamma(0)} \to \mathbb{R}^n_{\bar{\gamma}_0(0)}$ be an isometry such that

$$I(\sigma'(d)) = \bar{\sigma}'(d), \qquad I(\gamma_0'(0)) = \bar{\gamma}_0(0).$$

Since M is simply connected with non-positive curvature, exp is a diffeomorphism and we may apply Corollary 1.35. We get

$$\rho(\sigma(0), \gamma_0(l_0) \geq \rho(\bar{\sigma}(0), \bar{\gamma}_0(l_0)).$$

Thus after possibly increasing the angle between $\bar{\sigma}, \bar{\gamma}_0$, we obtain a hinge with distance between the endpoints equal to $\sigma(0), \gamma_0(l_0)$. Reasoning as in steps (1) and (2) in the proof of Toponogov's Theorem, we conclude that there exists a unique triangle $(\bar{\sigma}, \bar{\gamma}_0, \bar{\gamma}_1)$ with sides the same length as those of $(\sigma, \gamma_0, \gamma_1)$ and angles no smaller. In particular, $\beta + \alpha_1 + \delta \leq \pi$. Similarly $\hat{\beta} + \alpha_3 + \hat{\delta} \leq \pi$. Therefore,

$$\beta + \alpha_1 + \delta = \hat{\beta} + \alpha_3 + \hat{\delta} = \pi, \ \beta + \hat{\beta} = \alpha_0 \ \text{ and } \ \delta + \hat{\delta} = \alpha_2.$$

It follows that

$$\rho(\sigma(0), \gamma_0(l_0)) = \rho(\bar{\gamma}(0), \bar{\gamma}_0(l_0)).$$

Therefore

$$\exp_{\bar{\gamma}(0)} \circ I \circ \exp^{-1}_{\gamma_0(0)}(\gamma_2)$$

is the minimal segment from $\bar{\gamma}(0)$ to $\bar{\gamma}_2(l_0)$. All the inequalities in the proofs of Rauch I and Corollary 1.35 become equalities and the lemma follows in a straightforward manner by examining the proofs of those results. □

LEMMA 9.5. *Let g be an isometry of M.*

(1) If g generates a subgroup which acts properly discontinuously, then g is of finite order if and only if $\min \delta_g = 0$, *i.e. if g has a fixed point.*

(2) If g is semi-simple and of infinite order, then any g-invariant totally convex subset of C_g splits isometrically as $D \times \mathbb{R}$, where D is convex and each geodesic $d \times \mathbb{R}$ is invariant under g. The isometry g acts on $D \times \mathbb{R}$ as $(1, f)$, where f is a translation.

PROOF. (1) If $\min \delta_g = 0$, then g has a fixed point and is of finite order, since the group it generates is properly discontinuous. Conversely, if g is of finite order, let $m \in M$ and $m, g(m), \ldots, g^{n-1}(m)$ be the orbit of m. Let $d = \max \rho(m, g^i(m))$. Then for any i, $g^i(B_d(m))$ contains the orbit of m, and by Theorem 5.14 is totally convex. Thus $\bigcap_i g^i(\bar{B}_d(m))$ is a nonempty g-invariant compact totally convex set. Notice that by construction,

$$g^i(m) \in \partial(\bigcap_i g^i(B_d(m))),$$

unless $g(p) = p$, in which case we are done. The family $\mathfrak{F}$ of nonempty compact g-invariant totally convex sets is partially ordered by inclusion. By compactness, the intersection of a decreasing sequence of such sets is again in $\mathfrak{F}$, i.e. is in particular nonempty. It follows by Zorn's Lemma that $\mathfrak{F}$ contains minimal elements. If C is a minimal element and $p \in \operatorname{int} C$, then the above argument shows that unless $p = g(p)$ (equivalently $\{p\} = C$) we must have $p \in \partial C$. This is a contradiction.

(2) If $\min \delta_g = l > 0$ and $m \in C_g$, consider the minimal geodesic γ from m to $g(m)$ with $\gamma(0) = m$. We claim that g takes $N_0 = \gamma'(0)^\perp$ to $N_l = \gamma'(l)^\perp$. Otherwise for some $v \in N_0$, we have

$$\langle \mathrm{d}g(v), -\gamma'(l) \rangle > 0.$$

Then by the first variation formula, δ_g would be strictly decreasing on $\exp_x tv$, which contradicts $m \in C_g$. Now if

$$\mathrm{d}g(\gamma'(0)) = -\gamma'(l),$$

we would have

$$g(\gamma(l/2)) = \gamma(l/2).$$

Hence

$$dg(\gamma'(0)) = \gamma'(l)$$

and γ is invariant under g. Since g acts by translation on γ and $\gamma : (-\infty, \infty) \to M$ is an embedding (Theorem 1.39), it follows that g has infinite order. Suppose γ_1 *and* γ_2 are both invariant under g. Since g acts uniformly on γ_1, there exist $m_1 \in \gamma_1, m_2 \in \gamma_2$ such that $\rho(m_1, m_2) = \rho(\gamma_1, \gamma_2)$. If σ is the minimal geodesic from m_1 to m_2, then σ is perpendicular to γ_1, γ_2. It follows from Lemma 9.4 that if τ_1, τ_2 denote the segments of γ_1, γ_2 from m_1 to $g(m_1), m_2$ to $g(m_2)$ then $\tau_1, \sigma, \tau_2, g(\sigma)$ span a flat totally geodesic parallelogram P. The geodesic segments of P which are parallel to the γ_i lie on geodesics invariant under g.

It follows easily that the unit tangent vectors to the invariant geodesics form a parallel field on the totally geodesic interior of C_g. Therefore by the de Rham Decomposition Theorem the interior of C_g splits locally isometrically. However, the splitting is clearly global and extends to all of C_g is an obvious fashion. □

The next lemma is the last essential preliminary to the proof of Theorem 9.1.

LEMMA 9.6. *Let g be a semi-simple isometry of M. If $C \neq \emptyset$ is closed, convex and invariant under g, then $C \cap C_g \neq \emptyset$.*

PROOF. Assume first that g has no fixed point. Let $p \in C_g$, let q be a point on C for which $\rho(p, q) = \rho(p, C)$. Consider the geodesic quadrilateral with vertices $p, q, g(p), g(q)$, and sides γ_0 from p to $g(p)$, γ_1 from p, to q, $g(\gamma_1)$ from $g(p)$ to $g(q)$, and $\gamma_2 \subset C$ from q to $g(q)$. Now since γ_0 is part of an invariant line, the sum of the angles at p and $g(p)$ is π. Since

$$\rho(p, q) = \rho(g(p), g(q)) = \rho(p, C) = \rho(g(p), C),$$

and $\gamma_2 \subset C$, by the first variation formula each of the angles at $q, g(q)$ is $\geq \frac{1}{2}\pi$. Then by Lemma 9.4, the quadrilateral is flat and totally geodesic and the angles at $q, g(q)$ are right angles. Since $L[\gamma_1] = L[g(\gamma_1)]$, the quadrilateral must be a rectangle. Hence $L[\gamma_0] = L[\gamma_2]$, which implies $q \in C_g$. If g has a fixed point p, choose q as above. If $q \neq g(q)$, we obtain in the same way a geodesic triangle, the sum of whose angles is greater than π. By the argument of Lemma 9.4 this is impossible. □

PROOF OF THEOREM 9.1. As explained earlier, we are actually going to show something more than the statement of the theorem, namely that the set $C_\Sigma = \bigcap_{g \in \Sigma} C_g$ is nonvoid and splits isometrically as $D \times \mathbb{R}^k$. It will then follow that we may take $E = d \times \mathbb{R}^k$ for any $d \in D$.

(1) First suppose that Σ is abelian and the torsion subgroup $T \subset \Sigma$ is nonempty. Let $t \in T, t \neq 1$. Then C_t is totally convex by Lemma 9.3, and

dimC_t < dim$M$ because $t \neq 1$ (an isometry is determined by its differential at a single point). By Lemma 9.2(2), C_t is invariant under Σ, and by induction on dimension it follows that C_t has the properties claimed in the theorem relative to Σ/A, where A is the subgroup of elements which act as the identity on C_t. But by the proper discontinuity of Σ on M it follows that A is finite. This immediately implies that the theorem holds for Σ itself.

(2) Assume that Σ is abelian and without torsion. Let $g \in \Sigma$. By Lemma 9.2(2), C_g is invariant under Σ, and Σ acts properly discontinuously on C_g. By Lemma 9.6, the restriction of any g_1 to C_g is semi-simple. By Lemma 9.5, C_g splits isometrically as $D \times \mathbb{R}$ and g acts by $(1, h)$ where h is a translation. If γ is an invariant geodesic for g, then for any $g_1 \in \Sigma$,

$$g(g_1(\gamma)) = g_1(g(\gamma)) = g_1(\gamma).$$

Hence g_1 preserves the splitting. Since h is a translation and g_1 commutes with g, it is easily checked that the projection of g_1 on $\mathbb{R}$ is a translation. Let Σ/A denote the projection of Σ on D.

By Lemma 9.2(5,6), Σ/A acts properly discontinuously by semi-simple isometries on D. Hence by induction on dimension $C_{\Sigma/A} \subset D$ is nonempty and splits as $D' \times \mathbb{R}^{k-1}$, invariant under Σ/A where $\mathbb{R}^{k-1}$ is spanned by invariant lines. Moreover, the projection of Σ/A on $\mathbb{R}^{k-1}$ acts uniformly, properly discontinuously on $\mathbb{R}^{k-1}$ and each $g_1 \in \Sigma/A$ acts as $(1, f)$, where f is a translation. Then clearly each element of Σ acts as $(1, f, l)$ on

$$C_g = D' \times \mathbb{R}^{k-1} \times \mathbb{R}.$$

and $\mathbb{R}^{k-1} \times \mathbb{R} = \mathbb{R}^k$ is spanned by invariant lines. The projection of Σ on $\mathbb{R}^{k-1} \times \mathbb{R}$ is clearly uniform, properly discontinuous and without kernel. Taking $d \times \mathbb{R}^k = E$, this finishes the proof in case Σ is abelian.

(3) Now let

$$\Sigma = \Sigma^0 \supset \Sigma^1 \supset \cdots \supset \Sigma^n = e.$$

Then by induction we may assume that C_{Σ^1} is nonempty and splits as $D \times \mathbb{R}^l$, where Σ^1 acts as above. Since Σ^1 is normal in Σ, by (2) of Lemma 9.3, C_{Σ^1} is invariant under $\Sigma^0 = \Sigma$. Similarly, since each $d \times \mathbb{R}^l$ is spanned by the invariant lines through d of members of Σ^1, it follows that Σ preserves the splitting. Now Σ/Σ^1 is abelian and by the above, for the projection of Σ/Σ^1 on D, we have $C_{\Sigma/\Sigma^1} = D' \times \mathbb{R}^k$, where Σ/Σ^1 acts as above. Then clearly $C_\Sigma = D' \times \mathbb{R}^k \times \mathbb{R}^l$ and Σ acts as claimed. The kernel consists of just those elements which act as the identity on C_Σ. This is finite by proper discontinuity.

It follows from Lemma 9.2(2) that the hypothesis is satisfied if M is the universal covering space of a compact manifold. □

We derive some corollaries of Theorem 9.1.

COROLLARY 9.7. *Suppose that M is compact, $K_M \leq 0$, and $\pi_1(M) = \Gamma$ contains a solvable subgroup Σ. Then Σ is Bieberbach group and M contains the compact flat manifold E/Σ.*

PROOF. Lemma 9.3(1) implies Γ is semi-simple. Σ acts freely on the universal covering space $\widetilde{M}$, and hence by Lemma 9.5 is torsion-free. Hence Σ acts effectively on E as a Bieberbach group. □

Corollary 9.7 immediately implies:

COROLLARY 9.8. *$\pi_1(M)$ has a solvable subgroup of finite index if and only if M is flat.*

COROLLARY 9.9 (Preismann). *If M is compact and $K_M < 0$, then every abelian subgroup is infinite cyclic.*

PROOF. This actually follows from Lemma 9.5 and the nonexistence of a flat subspace. □

COROLLARY 9.10. *Let $\partial M = \emptyset$. Assume M/Γ compact, that Γ acts properly discontinuously, and let Σ be a solvable normal subgroup of Γ. Then $M = D \times E$ isometrically, where E is the Euclidean space corresponding to Σ in Theorem 9.1 and Γ preserves the product decomposition.*

PROOF. By the proof of Theorem 9.1, it suffices to show that C_Σ, which is nonvoid and splits as $D \times E$, is actually all of M. By Lemma 9.3(2), C_Σ is invariant under Γ. But we claim that since Γ acts uniformly, the convex hull $C(\Gamma(p))$ of the orbit of any point $p \in C_\Sigma$ is all of M. In fact, let $q \in C(\Gamma(p))$ and suppose that the tangent cone C_q at p is not M_q. Let γ be a normal geodesic such that $\gamma(0) = q$ and $\gamma'(0) \notin C_q$. Let $\alpha > 0$ be the minimum angle that γ makes with C_q. Suppose $M = \Gamma(B_d(q))$. Then for large t we have a trangle (σ, τ, γ) with τ the minimal segment from $\gamma(t)$ to some $g_t(q)$, σ the minimal segment from p to $g_t(q)$ with

$$L[\gamma] = t, \qquad L[\tau] < d, \qquad L[\sigma] \geq t - d.$$

Since M is simply connected, $K_M \leq 0$, and the angle at q is $\geq \alpha$, by comparison with Euclidean space such a triangle can not exist for t sufficiently large. □

Our final result provides another illustration of how algebraic properties of $\pi_1(M)$ are reflected in the geometry of M.

THEOREM 9.11. *Let M be compact, $K_M \leq 0$, and suppose $\pi_1(M) = \Gamma_1 \times \Gamma_2$ has no center. Then M splits isometrically as $M_1 \times M_2$ with $\pi_1(M_i) = \Gamma_i$.*

By use of Corollary 9.10, Theorem 9.11 may be suitably refined to include the case in which π_1 has nontrivial center.

Let $C(X)$ denote the convex hull of X.

LEMMA 9.12. *Let $\widetilde{M}$ be complete simply connected, and let $\Gamma_1 \times \Gamma_2$ be a group of isometries which acts uniformly on M. Let Γ_2 be finitely generated, act properly discontinuously and have no center. Then for all p, Γ_1 acts uniformly on $C(\Gamma_1(p))$.*

PROOF. We are going to show that if Γ_1 does not act uniformly on $C(\Gamma_1(p))$, we can construct a sequence, $g_i \in \Gamma_2$ of distinct elements with the property that for any $g \in \Gamma_2$, the sequence $\delta_{g_i^{-1} g g_i}(p)$ is uniformly bounded. Since Γ_2 acts discontinuously, it then follows for all sufficiently large distinct i and j, that

$$g_i^{-1} g g_i = g_j^{-1} g g_j.$$

Thus $g_j g_i^{-1}$ commutes with g. Letting g vary over a *finite* generating set of Γ_2, this implies that for sufficiently large distinct i and j, $g_j g_i^{-1}$ is in the center of Γ_2. This is a contradiction.

We now construct the sequence $\{g_i\}$. Since we assume that Γ_1 does not act uniformly on $C(\Gamma_1(p))$, there is a sequence $p_i \in C(\Gamma_1(p))$ such that $\rho(p_i, \Gamma_1(p)) \to \infty$. Because $\Gamma_1 \times \Gamma_2$ acts uniformly on $\widetilde{M}$, there exist sequences $\{f_i\} \subseteq \Gamma_1$, $\{g_i\} \subseteq \Gamma_2$ and a number d such that $\rho(f_i g_i(p), p_i) \leq d$. Then, by the triangle inequaity,

$$\begin{aligned}\rho(g_i(p), p) &= \rho(f_i g_i(p), f_i(p)) \\ &\geq \rho(p_i, f_i(p)) - \rho(p_i, f_i g_i(p)) \\ &\geq \rho(p_i, \Gamma_1(p)) - d.\end{aligned}$$

Therefore $\rho(p_i, \Gamma_1(p)) \to \infty$ implies $\rho(g_i(p), p) \to \infty$. Thus, after passing to a subsequence, we may assume that the $\{g_i\}$ are all distinct.

It remains to show that for given g, $\delta_{g_i^{-1} g g_i}(p)$ is uniformly bounded. We first note that if $\delta_g(p) = r$, then $\delta_g | C(\Gamma_1(p)) \leq r$. In fact since g centralizes Γ_1, C_g^r is invariant under Γ_1. By Lemma 9.3, C_g^r is convex. But $C(\Gamma_1(p))$ is the smallest convex set containing p which is invariant under Γ_1. Thus $C(\Gamma_1(p)) \subset C_g^r$, which is to say $\delta_g | C(\Gamma_1(p)) \leq r$. Now by the triangle inequality,

$$\begin{aligned}\delta_{g_i^{-1} g g_i}(p) &= \rho(g g_i(p), g_i(p)) \\ &\leq \rho(g g_i(p), g f_i^{-1}(p_i)) + \rho(g f_i^{-1}(p_i), f_i^{-1}(p_i)) \\ &\quad + \rho(f_i^{-1}(p_i), g_i(p)) \\ &= \rho(f_i g(p), p_i) + \delta_g(f_i^{-1}(p_i)) + \rho(p_i, f_i g_i(p)) \\ &\leq d + r + d.\end{aligned}$$

Thus the sequence $\{g_i\}$ has the desired property. □

PROOF OF THEOREM 9.11. Let $\widetilde{M}$ denote the universal covering space of M and $p \in M$. Let $\mathfrak{F}$ denote the family of nonempty closed Γ_1-invariant totally convex subsets of $C(\Gamma_1(p))$. By Lemma 9.12, there exists a compact set D such that

$$\Gamma_1(D) \supset C(\Gamma_1(p)).$$

It follows that for all $C \in \mathfrak{F}$, $C \cap D \neq \emptyset$. Moreover, if

$$C_1 \cap D \subseteq C_2 \cap D,$$

then for any $g \in \Gamma_2$,

$$C_1 \cap g(D) = g(C_1) \cap g(D) \subseteq g(C_2) \cap g(D) = C_2 \cap g(D).$$

Conversely, if $C_1 \subseteq C_2$, then $C_1 \cap D \subseteq C_2 \cap D$. This the sets $C \cap D$ (where $C \in \mathfrak{F}$) are partially ordered by inclusion. By compactness, the intersection of a decreasing sequence of such sets is nonempty. It follows therefore that the intersection of a decreasing sequence of members of $\mathfrak{F}$ is again a member of $\mathfrak{F}$. By Zorn's Lemma there exist minimal elements, say N_0, of $\mathfrak{F}$.

By Lemma 9.2(3), Γ_1 acts uniformly on N_0. By minimality of N_0, for any $p \in N_0$, we have $C(\Gamma_1(p)) = N_0$.

Let χ denote the collection of all Γ_1-invariant t.c.s. $N_1 \subset \widetilde{M}$ such that

$$\dim N_1 = \dim N_0,$$

Γ_1 acts uniformly, and for any $p \in N_1$ we have $C(\Gamma_1(p)) = N_1$. Because N_0, N_1 are Γ_1-invariant and Γ_1 acts uniformly, the set

$$K_0 = \{p \in N_0 | \rho(p, N_1) = \rho(N_0, N_1)\}$$

is nonempty. Let $p_0 \in K_0$ and $q_0 \in N_1$ such that

$$\rho(p_0, q_0) = \rho(N_0, N_1).$$

Let τ be the minimal geodesic from p_0 to q_0. Then by the first variation formula, τ makes an angle $\geq \frac{1}{2}\pi$ with any geodesic from p_0 which points into N_0. Then given any other pair p_1, q_1 such that

$$\rho(p_1, q_1) = \rho(N_0, N_1),$$

applying Lemma 9.4 to the quadrilateral determined by $\{p_0, q_0, p_1, q_1\}$ shows that K_0 is convex and $C(K_0 \cup K_1)$ splits isometrically as $K_0 \times I$ with any $K_0 \times t$ invariant under Γ_1. However, clearly $K_0 \in \chi$ so $K_0 = N_0$. In particular distinct members of χ are disjoint.

It also follows that the set of points which lie on some member of χ is convex. Clearly, however, this set is invariant under Γ and since Γ acts uniformly, by the proof of Corollary 9.10 it must be all of $\widetilde{M}$.

We now show that $\partial N = \emptyset$ for $N \in \chi$. Let $p \in \partial N$, $r \in \mathrm{int} N$ and γ the geodesic from r to p. Let q be a point on γ lying beyond p. Hence $q \notin N$ and therefore $q \in N_1 \in \chi$ for some $N_1 \neq N$. Since

$$C(N \cup N_1) = N \times I,$$

the triangle with vertices at r, p, q has two right angles, at r and p, which is impossible since $K \leq 0$. Thus $\partial N = \emptyset$.

Using the splitting $C(N \cup N_1) = N \times I$, it now follows easily that the tangent spaces to the members of χ define a smooth Γ-invariant *parallel* distribution on N. Thus by the de Rham Decomposition Theorem, $\widetilde{M}$ has a Γ-invariant isometric splitting $\widetilde{M} = \widetilde{M}_1 \times \widetilde{M}_2$ on which Γ_1 acts as $(f, 1)$.

We now show that the projection of the action of Γ_2 on M_1 is trivial. Let $g_2 \in \Gamma_2$. Then g_2 acts by (f_2, g_2). If f is any isometry of M_1 centralized by the uniform group Γ_1, by Lemma 9.2(6), δ_f is bounded. Since, by

Lemma 9.3, δ_f is convex, it is therefore constant. Hence by Lemma 9.5, either f_2 is the identity or it has infinite order. If $f_2 \neq 1$, we may assume that $f_2^n \notin \Gamma_1$ for all n because the center of Γ_1 is trivial. Then by Lemma 9.2(7), f_2 induces an infinite group I of nontrivial isometries of the compact manifold $\widetilde{M_1}/\Gamma_1$. Let T denote the closure of I in the (compact) group of isometries of M. Then T is compact, abelian, centralized by Γ_1 and its identity component T_0 is a torus. Let τ_t denote a 1-parameter circle subgroup of T_0 such that $\tau_{t_0} = 1$ but $\tau_t \neq 1$ for $0 < t < t_0$. τ_t generates a vector field

$$V_t(x) = \frac{d}{dt}\tau_t(x)$$

on M_1/Γ_1 which lifts to a vector field $\widetilde{V}$ on M_1. By integrating $\widetilde{V}$ we may lift τ_t continuously to a 1-parameter group $\widetilde{\tau}_t$ of isometries of M_1 such that $\widetilde{\tau}_{t_0} = 1$. Clearly for $0 < t < t_0$, $\widetilde{\tau}_t \neq 1$ and hence

$$\min \delta_{\tau_t} = a_t > 0.$$

By Lemma 9.5 there exists a Γ_1-invariant splitting $\widetilde{M_1} = D_t \times \mathbb{R}$ such that τ_t acts by $(1, f_t)$, where f_t is a translation by a real number $\bar{f}_t$. By considering small values of t and using continuity, one sees that the splitting $D_t \times \mathbb{R}$ is independent of t, and the factors $d \times \mathbb{R}$ are just the integral curves of $\widetilde{V}$. Similarly $t \to \bar{f}_t$ is a homomorphism. Alternatively, this could be deduced from the fact not proved here, that a Killing field on a compact manifold of non-positive curvature is parallel; see Bocher and Yano [1953]. Now $\tau_{t_0} = (1, f_{t_0})$ with $f_{t_0} \neq 1$ induces the identity on $\widetilde{M_1}/\Gamma_1$ and hence is a member of Γ_1. But this contradicts the triviality of the center of Γ_1.

We now have

$$M = \widetilde{M}/\Gamma = \widetilde{M_1}/\Gamma_1 \times \widetilde{M_2}/\Gamma_2 = M_1 \times M_2.$$

Since M is compact, it follows in particular that $\widetilde{M_2}/\Gamma_2$ is compact, and the proof is complete. □

Bibliography

M. BERGER

[1959] Variétés Riemanniennes à courbure positive, *Bull. Soc. Math. France* **87**.

[1960] Les variétés Riemanniennes $\frac{1}{4}$-pincées, *Ann. Scuola Norm. Sup. Pisa (III)* **14**(2).

[1960] Sur quelques variétés Riemanniennes suffisamment pincées, *Bull. Soc. Math. France* **88**.

[1961] Les variétés Riemanniennes homogènes normales simplement connexes à courbure strictment positive, *Ann. Scuola Norm. Sup. Pisa (III)* **15**(3).

[1961] Sur les variétés à courbure positive de diamètre minimum, *Comment. Math. Helv.* **35**(1).

[1962] Les variétés Riemanniennes dont la courbure satisfait certaines conditions, *Proc. Intern. Congr. Mathematicians.*

[1962] An extension of Rauch's metric comparison theorem and some applications, *Illinois J. Math.* **6**(4).

[1962] On the characteristic of positively pinched Riemannian manifolds, *Proc. Natl. Acad. Sci.* **48**(11).

R. BISHOP and R. CRITTENDEN

[1964] *Geometry of Manifolds* (Academic Press, New York).

S. BOCHNER and K. YANO

[1953] Curvature and Betti Numbers, *Ann. Math. Stud. (Princeton Univ. Press, Princeton, N.J.).*

A. BOREL

[1963] Compact Clifford-Klein forms of symmetric spaces, *Topology* **2**(2).

Yu.D. BURAGO and V.A. TOPONOGOV

[1973] On three-dimensional Riemannian manifolds whose curvature is bounded from above, *Mat. Zametki* **13** (in Russian).

J. CHEEGER

[1967] Comparison and finiteness theorems for Riemannian manifolds, Thesis Princeton Univ., Princeton, N.J.

[1969] Pinching theorems for a certain class of Riemannian manifolds, *Am. J. Math.* **91**(3).

[1970] Finiteness theorems for Riemannian manifolds, *Am. J. Math.* **92**(1).

[1973] Some examples of manifolds of nonnegative curvature, *J. Differential Geometry* **8**(9).

J. CHEEGER and D. GROMOLL

[1968] The structure of complete manifolds of nonnegative curvature, *Bull. Ann. Math. Soc.* **74**(6).

[1971] The splitting theorem for manifolds of nonnegative Ricci curvature, *J. Differential Geometry* **6**(1).

[1972] On the structure of complete manifolds of nonnegative curvature, *Ann. Math.* **96**(3), 413-443.

[1972] The injectivity radius of $\frac{1}{4}$-pinched Riemannian manifold, Reprint.

S. S. CHERN

[1944] A simple intrinsic proof of the Gauss-Bonnet formula for closed Riemannian manifolds, *Ann. Math.* **45**(2).

C. CHEVALLEY

[1946] Theory of Lie Groups (Princeton Univ. Press, Princeton, N.J.).

S. COHN-VOSSEN

[1935] Kürzeste Wege Totalkrümmung auf Flächen, *Comp. Math.* **2**, 69-133.

[1936] Totalkrümmung und geodätische Linien auf einfachzusammenhängenden offenen vollständigen Flächenstücken, *Rec. Math. Moscou* **43**, 139-163.

H. S. M. COXETER

[1957] Non-Euclidean Geometry (Univ. of Toronto Press, Toronto).

R. CRITTENDEN

[1962] Minimum and conjugate points in symmetric spaces, *Can. J. Math.* **14**, 320-328.

P. EBERLEIN

[1971] Manifolds admitting no metric of constant negative curvature, *J. Differential Geometry* **5**(1).

D. EBIN and J. MARSDEN

[1970] Groups of diffeomorphisms and the motion of an incompressible fluid, *Ann. Math.* **92**(1).

M. H. FREEDMAN

[1982] The topology of four-dimensional manifolds, *J. Differential Geometry* **173**, 357-453.

D. GROMOLL

[1966] Differenzierbare Strukturen und Metriken positiver Krümmung auf Spären, *Math. Ann.* **164**.

D. GROMOLL and W. MEYER

[1969] On complete open manifolds of positive curvature, *Ann. Math.* **90**(1).

[1974] An exotic sphere with non-negative sectional curvature, *Ann. Math.* **100**(1).

D. GROMOLL and J. WOLF

[1971] Some relations between the metric structure and the algebraic structure of the fundamental group in manifolds of nonpositive curvature, *Bull. Am. Math. Soc.* **77**(4).

S. HELGASON

[1962] Differential Geometry and Symmetric Spaces (Academic Press, New York).

N. HICKS

[1965] Notes on Differential Geometry (Van, Nostrand, New York, 1965).

N. J. HITCHIN

[1972] The space of harmonic spinors, Ph.D. Thesis, University of Oxford.

W. KLINGENBERG

[1959] Contributions to Riemannian geometry in the large, *Ann. Math* **69**.

[1961] Über Riemannische Mannigfaltigkeiten mit positiver Krümmung, *Comment. Math. Helv.* **35**.

[1963] Manifolds with restricted conjugate locus, *Ann. Math.* **75**(3).

[1969] Differentialgeometrie im Grossen, *Ber. Math. Forschungsinst. Oberwolfach.*

W. KLINGENBERG, D. GROMOLL and W. MEYER

[1968] Riemannische Geometrie im Grossen (Springer, Berlin).

S. KOBAYASHI and K. NOMIZU

[1963] Foundations of Differential Geometry, Vol. 1 (Interscience, New York).

[1969] Foundations of Differential Geometry, Vol. 2 (Interscience, New York).

H. B. LAWSON and S. T. YAU

[1972] On compact manifolds of nonpositive curvature, *J. Differential Geometry* **7**.

J. MILNOR

[1963] Morse Theory, Ann. Math. Stud. (Princeton Univ. Press, Princeton, N.J.).

[1968] A note on curvature and the fundamental group, *J. Differential Geometry* **2**.

M. MORSE

[1934] The Calculus of Variations in the Large (Am. Math. Soc., Providence, R.I.; reprinted 1966).

S. MEYERS and N. STEENROD

[1939] The group of isometries of a Riemannian manifold, *Ann. Math.* **40**.

K. NOMIZU

[1956] Lie Groups and Differential Geometry (Math. Soc. Japan, Tokyo).

B. O'NEILL

[1966] The fundamental equations of a submersion, *Michigan Math. J.* **13**.

R. S. PALAIS

[1969] The Morse Lemma for Banach spaces, *Bull. Am. Math. Soc.* **75**(5).

G. PERELMAN

[2002] The Entropy Formula for the Ricci Flow and its Geometric Applications, http://arXiv.org/abs/math.DG/0211159

[2003] Ricci Flow with Surgery on Three-Manifolds, http://arxiv.org/abs/math.DG/0303109

[2003] Finite Extinction Time for the Solutions to the Ricci Flow on Certain Three-Manifolds, http://arxiv.org/abs/math.DG/0307245

M. M. POSTNIKOV

[1965] The Variation Theory of Geodesics (Nauka, Moscow; in Russian).

H. E. RAUCH

[1951] A contribution to Riemannian geometry in the large, *Ann. Math.* **54**.

E. RUH

[1971] Curvature and differentiable structures on spheres, *Bull. Am. Math. Soc.* **77**(1).

J. T. SCHWARTZ

[1964] Nonlinear functional analysis, Notes, New York Univ., New York.

J. P. SERRE

[1951] Homologie singulière des espaces fibrés, *Ann. Math.* **54**.

Y. SHIKATA

[1967] On the differentiable pinching problem, *Osaka J. Math.* **4**(2).

S. SMALE

[1956] Generalized Poincaré conjecture in dimensions greater than 4, *Ann. Math.* **64**.

E. SPANIER

[1966] Algebraic Topology (McGraw-Hill, New York).

S. STERNBERG

[1964] Lectures of Differential Geometry (Prentice-Hall, Englewood Cliffs, N.J.).

M. SUGIMOTO, K. SHIOHAMA and H. KARCHER

[1971] On the differntiable pinching problem, *Math. Ann.* **195**.

V. A. TOPONOGOV

[1959] Riemann spaces with the curvature bounded below, *Uspehi Mat. Nauk* **14** (in Russian).

[1964] Spaces with straight lines, *Am. Math. Soc. Transl.* **37**.

Y. TSUKAMOTO

[1961] On Riemannian manifolds with positive curvature, *Mem. Fac. Sci. Kyushu Univ. (A)* **15**(2).

[1963] A proof of Berger's theorem, *Mem. Fac. Sci. Kyushu Univ. (A)* **17**(2).

[1966] On certain Riemannian manifolds of positive curvature, *Tohoku Math. J. (2)* **18**(1).

N. WALLACH

[1972] Three new examples of compact manifolds admitting Riemannian structures of positive curvature, *Bull. Am. Math. Soc.* **78**.

[1972] Homogeneous Riemannian manifolds of strictly positive curvature, *Ann. Math.* **75**.

F. WARNER

[1966] Extension of the Rauch comparison theorem to submanifolds, *Trans. Am. Math. Soc.*

A. WEINSTEIN

[1967] On the homotopy type of positively pinched manifolds. *Arch. Math.* **18**.

[1969] Symplectic structures on Banach manifolds, *Bull. Am. Math. Soc.* **75**(5).

J. H. C. WHITEHEAD

[1932] Convex regions in the geometry of paths, *Quart. J. Math. Oxford* **3**.

J. WOLF

[1966] Spaces of Constant Curvature (McGraw-Hill, New York).

Additional Bibliography

[1] Marcel Berger. Systoles et applications selon Gromov. *Astérisque*, (216):Exp. No. 771, 5, 279–310, 1993. Séminaire Bourbaki, Vol. 1992/93.

[2] Marcel Berger. *Riemannian geometry during the second half of the twentieth century*, volume 17 of *University Lecture Series*. American Mathematical Society, Providence, RI, 2000. Reprint of the 1998 original.

[3] Marcel Berger. *A panoramic view of Riemannian geometry*. Springer-Verlag, Berlin, 2003.

[4] Yu. Burago, M. Gromov, and G. Perel′man. A. D. Aleksandrov spaces with curvatures bounded below. *Uspekhi Mat. Nauk*, 47(2(284)):3–51, 222, 1992.

[5] Peter Buser and Hermann Karcher. *Gromov's almost flat manifolds*, volume 81 of *Astérisque*. Société Mathématique de France, Paris, 1981.

[6] Leonard S. Charlap. *Bieberbach groups and flat manifolds*. Universitext. Springer-Verlag, New York, 1986.

[7] J. Cheeger and K. Grove. *Metric and Comparison Geometry*. Surveys in Differential Geometry; XI. Internaitonal Press, Summerville, 2007.

[8] Jeff Cheeger. Critical points of distance functions and applications to geometry. In *Geometric topology: recent developments (Montecatini Terme, 1990)*, volume 1504 of *Lecture Notes in Math.*, pages 1–38. Springer, Berlin, 1991.

[9] Jeff Cheeger. *Degeneration of Riemannian metrics under Ricci curvature bounds*. Lezioni Fermiane. [Fermi Lectures]. Scuola Normale Superiore, Pisa, 2001.

[10] Jeff Cheeger. Degeneration of Einstein metrics and metrics with special holonomy. In *Surveys in differential geometry, Vol. VIII (Boston, MA, 2002)*, Surv. Differ. Geom., VIII, pages 29–73. Int. Press, Somerville, MA, 2003.

[11] Jeff Cheeger, Kenji Fukaya, and Mikhael Gromov. Nilpotent structures and invariant metrics on collapsed manifolds. *J. Amer. Math. Soc.*, 5(2):327–372, 1992.

[12] Sylvestre Gallot. Volumes, courbure de Ricci et convergence des variétés (d'après T. H. Colding et Cheeger-Colding). *Astérisque*, (252):Exp. No. 835, 3, 7–32, 1998. Séminaire Bourbaki. Vol. 1997/98.

[13] M. Gromov. Synthetic geometry in Riemannian manifolds. In *Proceedings of the International Congress of Mathematicians (Helsinki, 1978)*, pages 415–419, Helsinki, 1980. Acad. Sci. Fennica.

[14] M. Gromov. Hyperbolic groups. In *Essays in group theory*, volume 8 of *Math. Sci. Res. Inst. Publ.*, pages 75–263. Springer, New York, 1987.

[15] M. Gromov. Sign and geometric meaning of curvature. *Rend. Sem. Mat. Fis. Milano*, 61:9–123 (1994), 1991.

[16] M. Gromov and P. Pansu. Rigidity of lattices: an introduction. In *Geometric topology: recent developments (Montecatini Terme, 1990)*, volume 1504 of *Lecture Notes in Math.*, pages 39–137. Springer, Berlin, 1991.

[17] Mikhael Gromov. Carnot-Carathéodory spaces seen from within. In *Sub-Riemannian geometry*, volume 144 of *Progr. Math.*, pages 79–323. Birkhäuser, Basel, 1996.

[18] Misha Gromov. *Metric structures for Riemannian and non-Riemannian spaces*. Modern Birkhäuser Classics. Birkhäuser Boston Inc., Boston, MA, english edition, 2007.

Based on the 1981 French original, With appendices by M. Katz, P. Pansu and S. Semmes, Translated from the French by Sean Michael Bates.

[19] Karsten Grove. Critical point theory for distance functions. In *Differential geometry: Riemannian geometry (Los Angeles, CA, 1990)*, volume 54 of *Proc. Sympos. Pure Math.*, pages 357–385. Amer. Math. Soc., Providence, RI, 1993.

[20] Karsten Grove. *Riemannian geometry: a metric entrance*, volume 65 of *Lecture Notes Series (Aarhus)*. University of Aarhus, Department of Mathematics, Aarhus, 1999.

[21] Karsten Grove. Finiteness theorems in Riemannian geometry. In *Explorations in complex and Riemannian geometry*, volume 332 of *Contemp. Math.*, pages 101–120. Amer. Math. Soc., Providence, RI, 2003.

[22] Karsten Grove. Aspects of comparison geometry. In *Perspectives in Riemannian geometry*, volume 40 of *CRM Proc. Lecture Notes*, pages 157–181. Amer. Math. Soc., Providence, RI, 2006.

[23] P. Pansu. Effondrement des variétés riemanniennes, d'après J. Cheeger et M. Gromov. *Astérisque*, (121-122):63–82, 1985. Seminar Bourbaki, Vol. 1983/84.

[24] Peter Petersen. Convergence theorems in Riemannian geometry. In *Comparison geometry (Berkeley, CA, 1993–94)*, volume 30 of *Math. Sci. Res. Inst. Publ.*, pages 167–202. Cambridge Univ. Press, Cambridge, 1997.

[25] Peter Petersen. Rigidity and compactness of Einstein metrics. In *Surveys in differential geometry: essays on Einstein manifolds*, Surv. Differ. Geom., VI, pages 221–234. Int. Press, Boston, MA, 1999.

[26] Peter Petersen. *Riemannian geometry*, volume 171 of *Graduate Texts in Mathematics*. Springer, New York, second edition, 2006.

Index

About this book

The central theme of this book is the interaction between the curvature of a complete Riemannian manifold and its topology and global geometry.

The first five chapters are preparatory in nature. They begin with a very concise introduction to Riemannian geometry, followed by an exposition of Toponogov's theorem—the first such treatment in a book in English. Next comes a detailed presentation of homogeneous spaces in which the main goal is to find formulas for their curvature. A quick chapter of Morse theory is followed by one on the injectivity radius.

Chapters 6–9 deal with the most relevant contributions to the subject in the years 1959 to 1974. These include the pinching (or sphere) theorem, Berger's theorem for symmetric spaces, the differentiable sphere theorem, the structure of complete manifolds of non-negative curvature, and finally, results about the structure of complete manifolds of non-positive curvature. Emphasis is given to the phenomenon of rigidity, namely, the fact that new topological types can appear when assumptions on the curvature are relaxed from strict to weak inequalities, but only in a restricted way, which usually involves an isometry.

Much of the material, particularly the last four chapters, was essentially state-of-the-art when the book first appeared in 1975. Since then, the subject has exploded, but the material covered in the book still represents an essential prerequisite for anyone who wants to work in the field.